I0477474

Preface

Throughout history, humanity has sought to understand the fundamental truths of the universe—those timeless principles that govern everything from the smallest atom to the vastness of the cosmos. These truths, once discovered, have the power to transform not only our understanding but also the way we live, create, and build our future. The Unified Field Theory (UFT) offers such a framework: a profound synthesis of science, metaphysics, and philosophy that reveals the interconnectedness of all things.

But understanding is only the beginning. The true test of any theory lies in its ability to create tangible change. How can these universal principles be applied to the challenges we face today? What practical tools can they offer for solving global crises, advancing technology, fostering unity, and empowering individuals? This book, *Unified Harmony: Real-World Applications of the Unified Field Theory*, is a response to those questions.

In these pages, we explore how the principles of the UFT—rooted in harmony, proportionality, and interconnectedness—can guide humanity toward a balanced and sustainable future. From reimagining energy systems and governance structures to fostering personal growth and resilience, this work bridges the profound with the practical, offering a blueprint for aligning our actions with the universal rhythms that sustain life and progress.

This book is not just about theory. It is about action. It is an invitation to leaders, innovators, and individuals to embrace these principles and apply them in their daily lives, their communities, and their fields of expertise. The ideas here are meant to inspire, challenge, and empower—to show that a more harmonious world is not just possible but within reach when we align with the universal patterns that govern all things.

Whether you are a scientist, a visionary, or simply someone seeking a deeper connection to the world around you, this book offers tools and insights for creating meaningful change. Together, let us explore how the timeless truths of the UFT can transform the way we think, live, and build the future.

This journey is one of discovery and transformation, but above all, it is a call to action—a reminder that the power to shape a harmonious future lies within each of us.

Welcome to *Unified Harmony*.

Introduction: A New Vision of the Universe

The universe is vast, mysterious, and seemingly boundless. From the smallest subatomic particles to the grandest galactic structures, humanity's quest to understand the cosmos has unveiled profound truths—yet also exposed deep gaps in our knowledge. Two of the most pressing challenges in modern science are the enigmas of **dark matter** and **dark energy**, which seem to govern the behavior of the universe but remain shrouded in mystery. Alongside them, **black holes**, the enigmatic engines of spacetime, defy our understanding with their paradoxical nature as both cosmic destroyers and creators.

Our traditional frameworks, while groundbreaking, rely heavily on concepts that remain speculative or incomplete. Hidden particles, unobservable forces, and mathematical constructs have helped patch gaps in understanding, but they often leave us asking: *Are we missing something fundamental?* This book proposes that the answer lies in rethinking the universe's underlying principles. What if the cosmos isn't governed by chaos or randomness but by harmony, proportion, and balance? What if the Golden Ratio (ϕ), a mathematical constant celebrated for its presence in nature and art, holds the key to unlocking the mysteries of the universe?

The Etheric Phi Gravitation Formula: A Framework of Harmony

At the heart of this exploration is the **Etheric Phi Gravitation Formula (EPGF)**. This revolutionary framework integrates two profound ideas:

1. **Ether**: A reimagined concept of a permeating field that structures spacetime, influencing matter and energy.

2. **Phi (ϕ)**: The Golden Ratio, which governs natural patterns and proportions, offering a blueprint for cosmic harmony.

By merging these ideas, the EPGF reframes seemingly inexplicable phenomena—dark matter, dark energy, and black holes—as manifestations of a deeper, unified order. In this vision, the cosmos is not an enigma of fragmented forces but a symphony of interconnected patterns.

Physics and Cosmology: The First Steps

The journey begins here, in **Physics and Cosmology**, where we examine how the EPGF redefines our understanding of the universe's largest mysteries:

- **Dark Matter and Dark Energy**: Reimagined as effects of ϕ-modulated Ether, eliminating the need for invisible particles or forces.

- **Black Holes**: Seen as harmonized nodes within the cosmic fabric, shaped by proportional dynamics that align with natural laws.

This section lays the foundation for applying \phi and Ether principles to other domains, from quantum computing to ethical governance. It challenges us to see the universe not as a chaotic expanse but as a living, breathing structure shaped by simplicity and harmony.

A Unified Vision for Science and Humanity

This book is more than a theoretical treatise; it is a call to action. It invites physicists, mathematicians, philosophers, and visionaries to collaborate across disciplines, exploring how ancient principles of harmony can inform modern science and technology. It also offers practical applications that extend far beyond the laboratory—touching energy systems, societal structures, and even personal growth.

The journey to unify the physical, mathematical, and metaphysical begins here, with the foundational mysteries of the cosmos. As we explore these ideas, let us challenge the assumptions of fragmentation and complexity and embrace a new paradigm of universal interconnectedness. The answers we seek may already be written into the fabric of the universe, waiting for us to uncover their song.

Subsection 1: Addressing Dark Matter and Dark Energy

The Invisible Puzzle of the Cosmos

In the vast tapestry of the universe, dark matter and dark energy stand as enigmatic threads, weaving gravitational effects and cosmic acceleration that defy explanation. These unseen forces—comprising over 95% of the universe's total mass-energy—have perplexed scientists for decades.

1. **Dark Matter**:

   - Astronomers observe galaxies where stars on the outskirts move with inexplicable speed, defying Newtonian and Einsteinian gravitational predictions. This has led to the hypothesis of dark matter: an invisible, non-luminous substance that holds galaxies together like the glue of the cosmos.

   - Yet, after years of searching for dark matter particles like WIMPs (Weakly Interacting Massive Particles), scientists have come up empty-handed.

2. **Dark Energy**:

- Even more mysterious, dark energy is the theorized force driving the accelerating expansion of the universe. This "anti-gravity" effect appears to push galaxies apart, yet its nature remains speculative—a placeholder for a profound gap in our understanding of spacetime.

The reliance on unobservable entities reveals a deeper truth: our current frameworks may be missing a more elegant explanation. Could these cosmic phenomena instead emerge from natural laws we have yet to fully grasp?

The Etheric Phi Gravitation Formula: A New Paradigm

The **Etheric Phi Gravitation Formula (EPGF)** offers a revolutionary perspective. Rather than invoking hidden particles or forces, it proposes that the universe's structure is governed by the interaction of spacetime with an **Ether field**, modulated by the **Golden Ratio (\phi)**. This approach blends ancient principles of harmony with cutting-edge physics to reframe dark matter and dark energy as byproducts of a unified, observable mechanism.

1. **The Ether Field Reimagined**:

   - Ether, once dismissed as a relic of pre-relativity physics, is reinterpreted as a dynamic quantum field that permeates spacetime. Unlike traditional models, this Ether is structured by \phi, the mathematical constant renowned for its symmetry and proportionality in nature.

   - This field creates self-similar patterns across cosmic scales, influencing gravitational dynamics in ways current models cannot predict.

2. **Addressing Dark Matter**:

   - The EPGF suggests that the gravitational anomalies attributed to dark matter—such as the flat rotation curves of galaxies—are not caused by invisible particles. Instead, they emerge naturally from \phi-based variations in the Ether field's density.

   - **Example**: In a spiral galaxy, regions of higher Ether density enhance gravitational pull, balancing the rotation speeds of stars across the galaxy without invoking unseen mass.

3. **Addressing Dark Energy**:

   - Similarly, dark energy's repulsive force can be understood through the harmonic resonance of the Ether field on cosmic scales. This resonance generates an outward push, aligning with observations of accelerating expansion.

- **Example**: The universe's expansion becomes not a chaotic phenomenon but a harmonic ripple, echoing the self-similar growth patterns seen in nature, from spirals in shells to the branching of trees.

Transforming Cosmology Through Application

1. **Reinterpreting Observations**:

- Telescopes like the James Webb Space Telescope or gravitational wave detectors could be recalibrated to test for anomalies predicted by the EPGF. For instance:

  - Subtle shifts in galaxy cluster behavior.

  - Variations in gravitational lensing that align with ϕ-modulated models.

2. **Cosmological Simulations**:

- The integration of ϕ-modulated Ether dynamics into simulations could provide:

  - Accurate predictions for galaxy formation and motion.

  - Insights into the large-scale structure of the universe that reconcile current gaps in data.

3. **Guiding Experiments**:

- Instruments like the Vera Rubin Observatory, designed to measure dark energy, could test for correlations between ϕ-based patterns and the distribution of mass-energy across the cosmos.

A Cosmos of Harmony

The Etheric Phi Gravitation Formula invites us to reimagine the universe not as a chaotic interplay of mysterious entities but as a harmonious system guided by natural laws of proportion and balance. By shifting our perspective:

- **We bridge gaps**: Dark matter and dark energy are no longer unsolvable mysteries but manifestations of universal harmony.

- **We simplify science**: Elegant principles replace complex conjectures.

- **We inspire collaboration**: Uniting physicists, mathematicians, and cosmologists in a shared quest for a deeper understanding of the universe.

The EPGF transforms the question from "What is dark matter?" to "How does the universe achieve its balance?"—and with this shift, we open the door to a profound revolution in cosmology.

Subsection 2: Black Hole Dynamics

The Enigma of Black Holes

Black holes, the cosmic titans of spacetime, have long fascinated and mystified physicists. Defined as regions of extreme density where gravity becomes inescapable, black holes challenge our understanding of the universe on every front.

1. **Current Understanding**:

   - **Event Horizons**: The boundary beyond which nothing, not even light, can escape. A place where spacetime is so warped that traditional physics breaks down.

   - **Singularity**: Theoretical point at the center of a black hole where density is infinite, and the laws of physics cease to apply.

   - **Spin Dynamics**: Black holes rotate, dragging spacetime with them in a phenomenon known as frame-dragging. Their spin rates influence everything from accretion disks to jet emissions.

2. **Unresolved Questions**:

   - How do black holes interact with their surrounding environments at quantum and relativistic scales?

   - What governs their energy emissions, such as Hawking radiation or jets of relativistic particles?

   - Could they serve as portals to other regions of spacetime or dimensions?

These questions demand new frameworks that go beyond general relativity or quantum mechanics—a unified perspective that captures their duality as both cosmic destroyers and creators.

A New Lens: The Role of ϕ in Black Hole Dynamics

The **Etheric Phi Gravitation Formula (EPGF)** introduces ϕ-based symmetry to model black hole behavior. This approach reimagines black holes not as chaotic anomalies but as nodes of harmonic resonance within the Ether field.

1. ϕ-**Based Symmetry**:

- Black holes, in this framework, are structured by the same proportional relationships that define the natural world. Their spin, energy distributions, and event horizon behaviors are influenced by \phi-modulated spacetime.

- **Example**: The angular momentum of a spinning black hole could follow recursive patterns dictated by \phi, creating self-similar energy structures around the event horizon.

2. **Energy Emissions and Jets**:

- Black hole jets, which expel high-energy particles at near-light speeds, can be reframed as the result of \phi-based alignment in the Ether field.

- **Prediction**: The angle and intensity of jets may correlate with harmonic points of Ether resonance, providing a measurable signature of \phi dynamics.

3. **Event Horizons and Information Paradox**:

- The event horizon becomes a boundary where \phi-modulated Ether fields interact with incoming matter and radiation, influencing how information is stored or radiated.

- **Hypothesis**: \phi-based fields could provide a mechanism to resolve the black hole information paradox by encoding information in fractal-like patterns.

Future Directions for Black Hole Research

1. **Observational Validation**:

- Instruments like the Event Horizon Telescope (EHT), which produced the first image of a black hole, could be used to test for \phi-based anomalies in black hole spin rates, jet structures, or accretion disk patterns.

- **Expected Results**: Self-similar, recursive features around event horizons or jets consistent with \phi-modulated resonance.

2. **Simulating Black Hole Behavior**:

- Incorporating \phi-modulated dynamics into black hole simulations could yield new insights into:

  - The stability of accretion disks.

  - The formation and evolution of jets.

  - Energy exchange at the event horizon.

3. **Practical Implications**:

- Understanding \phi-based black hole dynamics could unlock new technologies, such as:

  - Advanced energy extraction methods from black hole spin (e.g., Penrose process optimizations).

  - Novel theories of spacetime manipulation for interstellar travel.

Implications for Unified Physics

By integrating \phi into black hole research, we gain:

- **Predictive Power**: A framework that anticipates observable patterns, linking cosmic phenomena to universal laws.

- **Harmonic Unity**: A perspective that aligns black holes with the natural harmony observed in other physical systems.

- **Practical Innovation**: New opportunities to harness black holes for technological advancement.

In the \phi-modulated universe, black holes are not chaotic voids but harmonized engines of cosmic evolution, revealing the intricate dance of matter, energy, and spacetime. The mysteries they hold are not barriers to understanding but invitations to explore the deeper symphony of the cosmos.

Conclusion for Section 1: Physics and Cosmology

Reimagining the Universe

From the invisible scaffolding of dark matter and energy to the gravitational powerhouses of black holes, the mysteries of the cosmos have long resisted comprehensive explanation. The **Etheric Phi Gravitation Formula (EPGF)** offers a radical yet elegant framework, reframing these phenomena as natural expressions of a harmonious universe structured by the Golden Ratio (\phi).

1. **Dark Matter and Dark Energy**:

   - The EPGF transforms these elusive concepts from hypothetical constructs into observable manifestations of a \phi-modulated Ether field.

   - This approach unifies galaxy dynamics, cosmic expansion, and gravitational anomalies under a single, coherent framework.

2. **Black Hole Dynamics**:

- By integrating \phi-based symmetry into black hole models, the EPGF reveals new insights into their spin, energy emissions, and event horizon behaviors.

- These findings not only enhance our understanding of black holes but also open pathways for practical applications, from advanced energy systems to interstellar exploration.

The Promise of Harmonic Science

The application of \phi and Ether principles marks a departure from fragmented theories reliant on unobservable entities. Instead, it highlights:

- **Interconnectedness**: A universe where phenomena resonate with natural laws of harmony and proportion.

- **Predictive Power**: A model capable of guiding observations, experiments, and simulations to uncover new cosmic truths.

- **Practical Potential**: Opportunities for innovation across physics, cosmology, and technology.

By uniting the metaphysical beauty of \phi with cutting-edge physics, the EPGF illuminates the intricate relationships that govern our universe, inviting collaboration across disciplines to explore its profound implications.

A Call to Action

The journey does not end here. To unlock the full potential of this paradigm, the scientific community must:

1. **Test Predictions**: Use advanced observational tools to validate \phi-modulated anomalies.

2. **Collaborate Across Fields**: Engage physicists, mathematicians, cosmologists, and technologists in a unified effort to expand the framework.

3. **Reimagine the Cosmos**: Embrace a perspective where simplicity and harmony replace complexity and speculation.

The EPGF redefines our understanding of the cosmos, not as a fragmented collection of phenomena but as a living symphony—one that we are only beginning to comprehend.

Introduction: Quantum Computing and the Harmony of Information

As humanity ventures further into the digital age, the power and potential of computing have transformed every aspect of modern life. Yet, classical computing faces limitations when addressing the complexities of nature, from molecular interactions to cosmological simulations. Enter **quantum computing**, a revolutionary technology that leverages the principles of quantum mechanics to process information in ways that were once unimaginable.

Quantum computing has already begun reshaping fields like cryptography, optimization, and artificial intelligence. However, its promise remains tethered to challenges—scalability, error correction, and a deeper understanding of quantum behavior. What if the solution to these challenges lies not in brute technological force but in the natural order of the universe itself?

This section explores how the **Golden Ratio (\phi)** and the **Etheric Phi Gravitation Formula (EPGF)** can guide the development of quantum computing. By integrating the self-similar, recursive nature of \phi into quantum systems, we can achieve greater efficiency, stability, and harmony in computation. Just as \phi creates order in nature's patterns, it may also provide the blueprint for optimizing quantum algorithms and architectures.

Quantum Computing Through the Lens of \phi

1. **A New Computational Paradigm**:

   - Classical computing processes information in linear bits—1s and 0s. Quantum computing, with qubits that exist in superposition, introduces exponential complexity. By aligning this complexity with \phi, we can bring order and efficiency to quantum systems.

2. **The Role of Harmony**:

   - \phi's inherent symmetry can guide the arrangement of qubits, the design of quantum gates, and the flow of information in quantum circuits.

3. **Practical Impact**:

   - From solving optimization problems to simulating complex biological systems, \phi-inspired quantum computing could unlock breakthroughs across industries.

Exploring the Subsections

This section will delve into:

- The potential for \phi-modulated structures in quantum systems to enhance stability and reduce error rates.

- Practical applications of \phi-inspired algorithms in fields like cryptography and data analysis.

- How the principles of harmony can make quantum computing more accessible and impactful for solving humanity's greatest challenges.

By weaving the order of \phi into the unpredictable realm of quantum mechanics, we can bridge the gap between technology's infinite possibilities and nature's timeless wisdom. Let us now explore how this harmony can redefine the future of computation.

Subsection 1: \phi-Modulated Structures in Quantum Systems

The Challenge of Quantum Complexity

Quantum computing, unlike classical computing, operates in the probabilistic realm of quantum mechanics. Its building blocks, **qubits**, leverage superposition and entanglement to process vast amounts of data simultaneously. While this gives quantum systems extraordinary power, it also introduces unique challenges:

- **Error Rates**: Quantum systems are highly sensitive to environmental noise, leading to decoherence and errors.

- **Scalability**: Maintaining stable quantum states across a large number of qubits is one of the most significant technological hurdles.

- **Optimization**: Designing quantum circuits and algorithms that efficiently solve problems without excessive overhead remains a complex task.

To overcome these challenges, a new approach is needed—one that aligns the inherent unpredictability of quantum systems with the natural order of the universe.

The Role of \phi in Stabilizing Quantum Systems

The **Golden Ratio (\phi)**, with its self-similar and recursive properties, offers a unique framework for addressing quantum instability and inefficiency.

1. **Harmonic Qubit Arrangements**:

   - Qubits in quantum systems must interact with precision to maintain coherence. By arranging qubits in \phi-**based geometric patterns**, such as spirals or fractals, their interactions become naturally harmonized, reducing interference.

- **Example**:

  - A quantum processor arranged using \phi-spirals minimizes cross-talk by leveraging proportional distances, optimizing qubit coherence.

2. **Circuit Design and Gate Operations**:

- Quantum gates, the building blocks of quantum circuits, operate on entangled qubits to perform computations. Integrating \phi-modulated timing and spacing into gate operations creates synchronized transitions, reducing error rates.

- **Prediction**:

  - Algorithms built with \phi-optimized gates could achieve faster, more accurate results compared to traditional designs.

3. **Resonance with Quantum Fields**:

- The **Etheric Phi Gravitation Formula (EPGF)** suggests that the Ether field, modulated by \phi, provides a stable framework for energy interactions. Embedding this principle in quantum systems could create more resilient qubit states, immune to external noise.

Applications of \phi-**Based Quantum Structures**

1. **Error Correction**:

- By structuring quantum error correction codes around \phi-inspired patterns, the system can detect and correct errors more efficiently, minimizing computational overhead.

2. **Quantum Simulations**:

- In simulating natural processes, such as molecular dynamics or climate systems, \phi-modulated structures align quantum simulations with the patterns inherent in the physical world, improving accuracy and energy efficiency.

3. **Scalable Architectures**:

- Hardware designed with \phi-based scalability principles can optimize the use of physical space while maintaining qubit coherence, paving the way for larger and more powerful quantum computers.

A Gateway to the Future

By embedding the principles of harmony and proportion into quantum systems, \phi-modulated structures transform quantum computing from an experimental frontier into a practical, reliable tool for innovation. These designs:

- **Reduce Complexity**: Natural patterns simplify the arrangement and interaction of qubits.

- **Enhance Stability**: Resonance with \phi-based harmonics improves coherence and error resistance.

- **Expand Potential**: Scalable, efficient architectures bring quantum computing closer to widespread application.

The integration of \phi into quantum systems is more than a technological advancement—it is a step toward aligning humanity's most advanced tools with the timeless wisdom of the cosmos.

Subsection 2: Practical Applications of \phi-Inspired Algorithms

The Rise of Quantum Algorithms

Quantum algorithms are at the heart of quantum computing's transformative potential. By leveraging the unique properties of qubits, these algorithms solve problems that classical computers would find intractable. Yet, their design and implementation face challenges of efficiency, stability, and adaptability.

To address these issues, \phi-**inspired algorithms** offer a new paradigm. Rooted in the principles of the Golden Ratio (\phi)—self-similarity, harmony, and proportionality—these algorithms align computational processes with the natural order of the universe. This approach enhances not only the efficiency of quantum systems but also their capacity to model real-world phenomena.

Designing Algorithms with \phi

1. **Self-Similarity in Recursive Processes**:

   - Many quantum algorithms rely on recursion, where a problem is broken down into smaller subproblems. Incorporating \phi-based recursion ensures that these subproblems are proportionally scaled, optimizing computational resources.

   - **Example**:

- Grover's algorithm for database search could be enhanced with \phi-modulated steps, reducing quantum gate operations while maintaining accuracy.

2. **Harmonic Timing in Quantum Gates**:

- The timing of quantum gate operations is critical for maintaining coherence. By structuring gate sequences based on \phi-derived intervals, algorithms achieve smoother transitions and reduced decoherence.

- **Prediction**:

  - Algorithms like Shor's factoring algorithm could process larger integers with greater stability using \phi-optimized gate timing.

3. **Dynamic Adaptation**:

- In quantum machine learning, adaptive algorithms adjust to new data inputs in real time. \phi-based proportional scaling can guide these adjustments, creating more responsive and efficient learning models.

Applications Across Industries

1. **Cryptography**:

- Quantum computers pose both risks and opportunities for cryptography. While they threaten traditional encryption methods, \phi-inspired algorithms can enhance post-quantum cryptography by optimizing key generation and security checks.

- **Example**:

  - A \phi-modulated encryption protocol could use recursive patterns to create keys that are more secure and computationally efficient.

2. **Optimization Problems**:

- Industries such as logistics, finance, and manufacturing face complex optimization challenges, from supply chain management to investment portfolios. \phi-inspired quantum algorithms can streamline these processes by aligning computational steps with natural efficiency.

- **Example**:

  - Quantum annealing systems designed with \phi-scaled energy landscapes could solve optimization problems faster and with lower error rates.

3. **Scientific Simulations**:

- Quantum algorithms excel at modeling complex systems, such as chemical reactions, climate patterns, or astrophysical phenomena. \phi-inspired algorithms, by mirroring the proportional patterns of nature, provide more accurate and efficient simulations.

- **Example**:

  - Simulating protein folding, a notoriously difficult computational problem, could benefit from \phi-guided energy minimization steps.

The Future of Algorithmic Harmony

\phi-inspired algorithms are not just theoretical constructs—they represent a transformative step in the evolution of quantum computing. By aligning computational processes with the universal principles of harmony, these algorithms:

- **Enhance Efficiency**: Fewer resources are required to achieve the same or greater computational power.

- **Improve Accuracy**: Natural patterns reduce errors and increase the fidelity of results.

- **Expand Accessibility**: Optimized algorithms lower the barriers to entry for industries and researchers adopting quantum technologies.

As quantum computing continues to evolve, \phi-inspired algorithms illuminate a path forward where nature's blueprint shapes the cutting edge of technology. They are not just tools for computation—they are manifestations of the timeless order embedded in the universe.

Subsection 3: Bridging Quantum and Classical Computing with \phi

The Divide Between Quantum and Classical Computing

Quantum and classical computing operate in fundamentally different realms. Classical systems process binary information through deterministic logic, while quantum systems exploit superposition and entanglement to perform computations that defy classical limits. Despite their distinct natures, a symbiotic relationship between these two paradigms is essential for unlocking the full potential of computational technology.

- **Challenges in Integration**:

- **Data Translation**: Translating classical data into quantum states and vice versa introduces inefficiencies.

- **Resource Bottlenecks**: Coordinating tasks between quantum processors and classical systems requires significant overhead.

- **Algorithmic Gaps**: Hybrid systems often struggle to reconcile quantum speed-ups with classical computational constraints.

Addressing these challenges requires a unifying principle that optimizes the flow of information between quantum and classical domains. Here, the **Golden Ratio (\phi)** emerges as a natural bridge, offering proportionality, efficiency, and harmony to integrate these two worlds.

How \phi Facilitates Integration

\phi-based principles provide a framework for designing hybrid systems where classical and quantum components work seamlessly together.

1. **Proportional Resource Allocation**:

   - By applying \phi-based scaling, computational resources can be allocated efficiently between quantum and classical processors.

   - **Example**:

     - A hybrid algorithm could assign \phi-proportioned workloads to quantum and classical units, ensuring that neither system becomes a bottleneck.

2. **Optimized Data Transfer**:

   - Data transfer between quantum and classical systems often involves high latency. Structuring these transfers around \phi-timed intervals reduces inefficiencies.

   - **Prediction**:

     - \phi-modulated timing could minimize qubit-to-bit conversion errors, enhancing the fidelity of hybrid computations.

3. **Algorithmic Harmony**:

   - Algorithms that require both quantum and classical processing can benefit from \phi-inspired recursive structures, ensuring smooth transitions between computational paradigms.

- **Example**:

  - A hybrid optimization algorithm could dynamically switch between quantum exploration and classical refinement phases, guided by \phi-scaled thresholds.

Applications of Hybrid Systems

1. **AI and Machine Learning**:

   - Quantum systems excel at exploring vast solution spaces, while classical systems are adept at refining results. \phi-based hybrid systems could balance these strengths, creating more robust and efficient AI models.

   - **Example**:

     - A \phi-guided machine learning framework could use quantum processors for feature selection and classical systems for model training.

2. **Real-Time Decision-Making**:

   - Industries such as finance and logistics require rapid computations that blend quantum speed with classical reliability. \phi-optimized hybrid algorithms could enhance decision-making in real time.

   - **Example**:

     - A stock trading system could leverage quantum systems for market predictions and classical processors for executing trades, harmonized by \phi-based timing.

3. **Simulation and Modeling**:

   - Many scientific problems require the computational power of quantum systems and the numerical stability of classical methods. \phi-modulated hybrid systems could unlock breakthroughs in fields like climate modeling and drug discovery.

   - **Example**:

     - A \phi-inspired simulation framework could dynamically allocate quantum resources for complex molecular interactions while using classical solvers for broader system behaviors.

The Future of Hybrid Computing

By bridging quantum and classical computing, \phi-based systems pave the way for a new era of computational harmony. These systems:

- **Integrate Strengths**: Leverage the best features of both paradigms.

- **Enhance Efficiency**: Reduce resource bottlenecks and computational overhead.

- **Expand Applications**: Make hybrid computing accessible to a wider range of industries and research fields.

In a world where quantum and classical computing increasingly intersect, \phi provides the blueprint for seamless integration, ensuring that technology advances in alignment with the natural rhythms of the universe.

Conclusion: Quantum Harmony for a New Era

The world of quantum computing represents the frontier of technological innovation, where the possibilities seem boundless, yet the challenges remain formidable. By integrating the principles of the **Golden Ratio (\phi)** into quantum systems, we uncover a path to address these challenges with elegance, efficiency, and harmony.

Key Insights from Section 2

1. \phi-**Modulated Structures**:

   - Quantum systems often struggle with stability and scalability. By arranging qubits, circuits, and energy flows in \phi-inspired patterns, we create architectures that naturally reduce errors and enhance coherence.

2. **Practical Algorithms**:

   - Quantum algorithms optimized with \phi-based principles provide superior performance in cryptography, optimization, and simulations. These algorithms leverage nature's inherent efficiency to achieve greater accuracy and speed.

3. **Bridging the Divide**:

   - The integration of quantum and classical computing remains essential for realizing the full potential of hybrid systems. \phi-inspired frameworks enable these systems to work in harmony, unlocking new applications in artificial intelligence, real-time decision-making, and scientific modeling.

A Blueprint for Future Innovation

As quantum computing continues to evolve, \phi-based principles serve as more than just a theoretical curiosity—they offer practical solutions for advancing technology in alignment with the universal laws of harmony and balance.

- **For Researchers**:

  - Quantum physicists and computer scientists can explore \phi-optimized designs and algorithms, testing their scalability and efficiency in real-world applications.

- **For Innovators**:

  - Industries on the cutting edge of AI, logistics, and cryptography can adopt \phi-based systems to enhance their computational power while reducing complexity.

- **For Society**:

  - Accessible quantum systems, optimized by \phi, democratize technology, making its benefits available across diverse sectors and communities.

A Harmonious Vision for Computation

Quantum computing has often been portrayed as a chaotic frontier, full of untapped power and untamed potential. By applying the timeless order of \phi, we move beyond this chaos into a realm of structured possibility, where the universe's natural rhythms guide humanity's most advanced tools.

This vision extends far beyond the realm of computation. It sets the stage for a world where technology evolves not as a force separate from nature but as a reflection of its most profound truths. As we continue this journey, let us embrace the harmony of \phi as a beacon for innovation, discovery, and the unification of human ingenuity with the cosmos itself.

Introduction: Energy Systems and the Rhythm of the Cosmos

Energy is the lifeblood of human progress. From ancient fires to modern power grids, our ability to harness and utilize energy has defined the trajectory of civilization. Yet, as humanity faces the dual crises of climate change and resource scarcity, the demand for more efficient, sustainable, and innovative energy systems has never been greater.

Traditional energy paradigms, while revolutionary in their time, often operate at odds with the natural world, leading to inefficiencies and environmental harm. What if our approach to energy could be transformed by aligning it with the fundamental principles of the universe? By integrating the harmony of the **Golden Ratio (ϕ)** and the **Etheric Phi Gravitation Formula (EPGF)**, we can unlock energy systems that are not only more efficient but also more attuned to the rhythms of nature.

This section explores how the principles of ϕ can guide advancements in energy generation, transmission, and storage, creating systems that mirror the elegance and balance of the cosmos. From renewable energy technologies to innovative transmission methods, ϕ-inspired designs offer solutions to the most pressing energy challenges of our time.

Energy Through the Lens of Harmony

1. **Renewable Energy Optimization**:

   - Harnessing the self-similar patterns of ϕ in solar panels, wind turbines, and wave energy systems can maximize efficiency and minimize waste.

2. **Energy Transmission**:

   - Leveraging the resonance of ϕ-modulated Ether fields to reduce energy loss in long-distance transmission.

3. **Sustainable Innovation**:

   - Designing energy systems that align with the natural rhythms of the Earth, fostering a sustainable and harmonious relationship with the environment.

Exploring the Subsections

This section will delve into:

- The application of ϕ-based principles in renewable energy systems, including solar, wind, and wave technologies.

- The role of ϕ-modulated Ether fields in revolutionizing energy transmission and reducing inefficiencies.

- The development of sustainable energy storage systems that utilize ϕ-inspired dynamics for greater capacity and longevity.

As we reimagine energy systems through the lens of harmony, we step closer to a future where human ingenuity and the natural world are not in opposition but in perfect balance.

Subsection 1: \phi-Optimized Renewable Energy Systems

The Challenges of Renewable Energy

Renewable energy technologies—solar, wind, and wave power—represent humanity's best hope for transitioning to a sustainable energy future. However, these systems are not without challenges:

1. **Efficiency**: Solar panels and wind turbines often fail to capture the full potential of available energy due to design and placement inefficiencies.

2. **Space Utilization**: Large-scale energy farms require significant land or ocean area, often creating environmental and logistical conflicts.

3. **Intermittency**: The variable nature of sunlight, wind, and waves makes consistent energy generation difficult.

To overcome these limitations, renewable systems must evolve beyond their current designs. The Golden Ratio (\phi) offers a pathway to optimize these technologies by aligning their structures and operations with nature's inherent efficiency.

Applying \phi to Renewable Energy

The self-similar and proportional qualities of \phi provide a blueprint for enhancing renewable energy systems.

1. **Solar Panels**:

   - Traditional solar panel arrays are often arranged in uniform grids, which fail to maximize sunlight capture throughout the day.

   - \phi-**Inspired Design**:

     - Panels arranged in \phi-**spiral patterns** can better track the sun's movement, increasing efficiency by minimizing shading and optimizing light capture.

   - **Example**:

     - A solar farm in a desert region could use \phi-spiral layouts to enhance energy output while reducing land usage by up to 30%.

2. **Wind Turbines**:

- Wind farms often experience turbulence caused by improperly spaced turbines, reducing overall efficiency.

- \phi-**Proportioned Spacing**:

  - Placing turbines at distances derived from \phi minimizes airflow interference and maximizes wind capture.

- **Prediction**:

  - A wind farm utilizing \phi-spacing could achieve a 15-20% increase in energy generation compared to conventional layouts.

3. **Wave Energy Systems**:

- Wave energy converters (WECs) must harness irregular and dynamic wave patterns, making placement and design critical.

- **Harmonic Structures**:

  - WECs arranged in \phi-**modulated clusters** align with natural wave frequencies, improving energy capture and reducing mechanical stress.

- **Example**:

  - A coastal wave farm could use \phi-aligned buoy systems to increase energy yield while extending device lifespan.

Benefits of \phi-Optimized Designs

1. **Increased Efficiency**:

- By mirroring natural patterns, \phi-based designs enhance energy capture and reduce losses.

2. **Reduced Footprint**:

- Optimized layouts require less land or water area, mitigating environmental impact and easing deployment.

3. **Harmonized Systems**:

- Aligning renewable technologies with the natural flow of energy creates systems that are not only more efficient but also more sustainable.

Real-World Applications and Future Directions

1. **Case Studies**:

   • Pilot projects incorporating \phi-based designs could demonstrate the feasibility and advantages of these systems.

   • **Example**: A solar farm in California adopts \phi-spiral layouts, achieving a 25% efficiency gain over traditional configurations.

2. **Collaborative Research**:

   • Engineers, architects, and environmental scientists can work together to refine \phi-inspired designs for different climates and geographies.

3. **Scaling Solutions**:

   • As \phi-optimized systems prove their value, they could be deployed at scale, revolutionizing the global renewable energy landscape.

A Harmonious Future

The integration of \phi into renewable energy systems is not just a technological innovation—it is a philosophical shift. By aligning human ingenuity with the inherent efficiency of nature, we can create energy systems that are both powerful and sustainable. In this vision, solar panels track the sun like spiraling sunflowers, wind turbines dance in harmonious patterns, and wave energy systems move in rhythm with the ocean's tides—a future where technology and nature work as one.

Subsection 2: \phi-Modulated Energy Transmission

The Inefficiencies of Traditional Transmission Systems

Energy transmission, the process of delivering electricity from its generation source to end users, is fraught with inefficiencies:

1. **Energy Loss**: Long-distance transmission often results in significant energy loss due to resistance in power lines.

2. **Infrastructure Costs**: Traditional grid systems require extensive and costly infrastructure to maintain stability.

3. **Environmental Impact**: Expanding transmission networks often disrupts ecosystems and landscapes.

Despite advancements in renewable energy, the benefits of clean generation are undermined by these systemic inefficiencies. To truly revolutionize energy distribution, transmission systems must embrace principles that enhance efficiency and sustainability. Enter \phi-**modulated transmission**, a framework inspired by the harmonic and proportional nature of the Golden Ratio (\phi).

Harnessing \phi in Energy Transmission

\phi-modulated transmission systems integrate the proportionality and resonance of \phi into energy distribution networks, optimizing the flow of electricity and reducing losses.

1. **Resonant Power Lines**:

 • Traditional transmission lines experience resistance and heat loss. By aligning the spacing and configuration of conductors with \phi-based proportions, energy flow becomes more harmonic, minimizing resistance.

 • **Example**:

   • A high-voltage transmission line designed with \phi-proportioned spacing reduces energy losses by up to 15%.

2. **Waveform Modulation**:

 • Electricity is transmitted as an alternating current (AC), which can experience phase mismatches over long distances. Modulating the waveform using \phi-based frequencies ensures smoother transitions and less energy dissipation.

 • **Prediction**:

   • \phi-modulated waveforms could improve energy efficiency by synchronizing power flows with the natural harmonics of the transmission system.

3. **Etheric Phi Fields**:

 • According to the **Etheric Phi Gravitation Formula (EPGF)**, \phi-modulated Ether fields can influence energy interactions. Embedding transmission systems within these fields could stabilize energy flows and reduce external interference.

 • **Hypothesis**:

   • Integrating EPGF principles into transmission infrastructure could create self-sustaining energy networks with near-zero losses.

Applications in Modern Energy Grids

1. **Microgrids**:

- Decentralized energy systems, or microgrids, benefit from \phi-optimized designs by harmonizing energy flow between localized generation sources and users.

  - **Example**:

    - A microgrid in a remote community uses \phi-proportioned conductor layouts to ensure stable, efficient energy delivery.

2. **Smart Grids**:

- Smart grids rely on real-time data to balance supply and demand. Incorporating \phi-based algorithms into grid management systems enhances predictive accuracy and operational efficiency.

  - **Example**:

    - A smart grid adjusts power distribution in \phi-modulated intervals, preventing overloads and reducing peak demand losses.

3. **Global Energy Networks**:

- Large-scale energy networks, such as those connecting renewable energy hubs across continents, require robust transmission systems. \phi-aligned designs ensure stability over vast distances, reducing the environmental footprint of infrastructure.

  - **Example**:

    - A transcontinental energy grid uses \phi-optimized transmission lines to deliver solar power from deserts to urban centers with minimal loss.

Realizing the Potential of \phi-Modulated Systems

To implement \phi-modulated transmission, collaboration across engineering, physics, and environmental science is essential:

1. **Pilot Projects**:

- Demonstrating the benefits of \phi-based transmission in localized systems paves the way for broader adoption.

- **Example**: A \phi-modulated grid in a renewable energy hub shows a measurable reduction in energy losses.

2. **Advanced Materials**:

- Developing superconductors and conductive materials aligned with \phi-based principles enhances the feasibility of large-scale systems.

3. **Policy Support**:

- Governments and energy providers must invest in research and infrastructure to enable the transition to \phi-modulated networks.

A Vision of Seamless Energy Flow

Imagine a world where energy flows like water through a river, unimpeded and harmonious. With \phi-modulated transmission, this vision becomes a reality. By aligning power grids with the natural rhythms of the universe, we not only improve efficiency but also foster a sustainable relationship with the environment. In this future, the energy we harness and distribute reflects the harmony of the cosmos itself.

Subsection 3: Sustainable Energy Storage with \phi

The Bottleneck of Energy Storage

Energy storage is the cornerstone of a renewable future. It bridges the gap between energy generation and consumption, ensuring that power is available when and where it's needed. However, the current landscape of energy storage faces significant hurdles:

1. **Capacity Limits**: Batteries and other storage systems struggle to store large amounts of energy efficiently.

2. **Longevity**: Frequent charge and discharge cycles degrade storage systems over time, reducing their lifespan.

3. **Environmental Concerns**: Many storage technologies rely on materials and processes that are environmentally harmful.

To overcome these challenges, energy storage systems must embrace principles that enhance capacity, durability, and sustainability. The Golden Ratio (\phi) provides a novel framework for designing storage solutions that align with nature's inherent efficiency and harmony.

The Role of \phi in Energy Storage

By integrating \phi-based principles into storage systems, we can optimize their design, operation, and scalability.

1. \phi-**Optimized Structures**:

• The internal arrangement of storage systems, such as battery electrodes or capacitors, plays a critical role in determining their efficiency. Using \phi-inspired fractal and spiral geometries improves energy density and flow.

  • **Example**:

    • A lithium-ion battery with \phi-designed electrode patterns reduces internal resistance, increasing both capacity and charge speed.

2. **Resonant Charging Cycles**:

• The charge and discharge cycles of storage systems often cause wear and inefficiency. By synchronizing these cycles with \phi-modulated intervals, systems experience smoother energy flows and reduced degradation.

  • **Prediction**:

    • Batteries employing \phi-aligned charging protocols could extend their lifespan by 20-30%.

3. **Etheric Energy Stabilization**:

• The **Etheric Phi Gravitation Formula (EPGF)** suggests that \phi-modulated Ether fields can stabilize energy interactions. Embedding storage systems within these fields could minimize energy loss and thermal stress.

  • **Hypothesis**:

    • \phi-stabilized storage systems could achieve near-perfect efficiency, transforming how we store and use energy.

Applications of \phi-Inspired Storage Systems

1. **Grid-Scale Storage**:

• Large-scale energy storage systems, such as those used for balancing renewable energy supply, benefit from \phi-optimized layouts and materials.

  • **Example**:

    • A grid-scale battery farm employing \phi-based designs achieves higher capacity and stability, supporting solar and wind energy integration.

2. **Portable Energy**:

- Portable storage solutions, like those in electric vehicles (EVs) or consumer electronics, require compact and efficient designs. \phi-inspired geometries enhance energy density while reducing weight.

  - **Example**:

    - An EV battery with \phi-optimized electrodes delivers longer range and faster charging compared to conventional designs.

3. **Sustainable Materials**:

- Incorporating \phi-aligned molecular structures into advanced materials, such as solid-state batteries, improves performance while reducing reliance on environmentally harmful elements like cobalt.

  - **Example**:

    - A solid-state battery using \phi-structured ceramics offers both higher safety and greater efficiency.

A Path Toward Sustainable Innovation

1. **Pilot Deployments**:

- Demonstrating \phi-inspired storage systems in renewable energy hubs or EV prototypes can validate their potential.

- **Example**: A renewable energy farm uses \phi-optimized batteries to store surplus solar and wind energy, achieving 30% greater efficiency.

2. **Collaborative Research**:

- Material scientists, energy engineers, and physicists must collaborate to refine \phi-based designs for diverse applications.

3. **Global Adoption**:

- Governments and industries must invest in \phi-inspired solutions to accelerate the transition to sustainable energy storage.

A Future of Endless Possibility

With \phi-inspired energy storage systems, humanity can transcend the limitations of today's technology. These designs not only enhance performance but also embody the principles of balance and sustainability. In this vision, energy is stored not as a finite resource but as a reflection of the universe's infinite harmony—a resource that works with nature, not against it.

Conclusion: Energy Systems Aligned with Universal Harmony

Energy is not merely a resource—it is the pulse of human civilization and the essence of the natural world. By embracing the principles of the **Golden Ratio (\phi)** and the **Etheric Phi Gravitation Formula (EPGF)**, we can reimagine energy systems that are not only efficient but also harmonious with the rhythms of the universe.

Key Insights from Section 3

1. **Renewable Energy Optimization**:

   • \phi-based designs enhance the efficiency and sustainability of solar, wind, and wave energy systems. These innovations align energy generation with nature's inherent patterns, reducing waste and maximizing potential.

2. **Energy Transmission**:

   • Long-distance power delivery is revolutionized by \phi-modulated transmission systems, which minimize losses and harmonize energy flows. This approach creates grids that are both resilient and environmentally conscious.

3. **Sustainable Storage**:

   • Energy storage technologies, inspired by \phi-based geometries and charging protocols, achieve greater capacity, longevity, and sustainability. These systems are poised to support the global transition to renewable energy on a massive scale.

A Vision for the Future

The integration of \phi into energy systems is not just a technological advancement—it is a philosophical shift. It challenges humanity to move beyond exploitative practices and toward a relationship with energy that is cooperative and regenerative.

1. **Efficiency Meets Sustainability**:

   • \phi-aligned systems demonstrate that efficiency and sustainability are not mutually exclusive. By mimicking the patterns of the cosmos, we can achieve both simultaneously.

2. **Harmonizing Technology and Nature**:

- In this new paradigm, energy systems are no longer rigid, wasteful constructs. They become dynamic and adaptive, reflecting the natural order that governs all life.

3. **Empowering Global Transformation**:

- From rural communities to urban megacities, \phi-inspired energy systems have the potential to democratize access to clean, reliable power. This global transformation bridges technological innovation with environmental stewardship.

A Call to Action

The journey to create energy systems aligned with universal harmony requires collaboration, imagination, and commitment:

- **For Innovators**: Explore how \phi-based designs can optimize generation, transmission, and storage technologies.

- **For Policymakers**: Support research and development in harmonic energy solutions to accelerate their deployment.

- **For Communities**: Advocate for energy systems that prioritize sustainability and harmony over short-term gains.

As humanity faces the dual crises of climate change and energy scarcity, the principles of \phi provide a beacon of hope. By aligning our energy systems with the timeless rhythms of the universe, we take a critical step toward a future where progress and preservation go hand in hand—a future powered by the boundless harmony of the cosmos.

Introduction: Societal Systems in Harmony with Universal Truths

Throughout history, humanity has grappled with the challenge of building societies that balance progress with fairness, innovation with sustainability, and individuality with collective well-being. Today, this challenge has grown more urgent than ever. From economic inequality and environmental crises to political polarization and technological upheaval, the systems that govern our world are often at odds with the natural harmony of the universe.

This section explores how the principles of the **Golden Ratio (**\phi**)** and the **Etheric Phi Gravitation Formula (EPGF)** can inspire the reimagining of societal systems. By aligning governance, economics, and community structures with the universal laws of balance and proportion, we can foster societies that are not only functional but also deeply harmonious.

Society Through the Lens of \phi

1. **Governance**:

   - Incorporating proportional representation and decision-making structures based on \phi, fostering more balanced leadership.

2. **Economics**:

   - Designing economic systems that reflect the fractal and self-similar growth patterns found in nature, reducing inequality and enhancing sustainability.

3. **Community Development**:

   - Creating urban and rural spaces that prioritize harmony between human needs and environmental stewardship.

Exploring the Subsections

This section will address:

- How \phi-based governance structures can lead to more balanced and representative leadership.

- The application of \phi in economic systems to create equitable and resilient markets.

- The role of harmonic design in urban planning and community development to foster well-being and sustainability.

As we apply the principles of \phi to societal systems, we uncover a vision of the world where human ingenuity and natural harmony coalesce—a world where systems evolve not through conflict but through cooperation, balance, and shared purpose.

Subsection 1: Governance Aligned with \phi

The Challenges of Modern Governance

Governance is the backbone of human civilization, yet it often struggles to reflect the balance and equity necessary for long-term stability:

1. **Imbalance of Power**: Many governance structures centralize power in ways that marginalize significant portions of the population.

2. **Inefficiency**: Bureaucratic systems often fail to adapt to changing circumstances, leading to stagnation or mismanagement.

3. **Polarization**: A lack of proportional representation fosters division, undermining collective progress.

These challenges call for a new approach to governance—one rooted in the universal principles of balance and harmony exemplified by the **Golden Ratio (\phi)**.

The Role of \phi in Governance

The Golden Ratio provides a natural blueprint for designing governance systems that reflect proportionality, fairness, and adaptability.

1. **Proportional Representation**:

   - \phi-inspired governance structures prioritize proportionality, ensuring that power is distributed in a way that reflects the diversity of the population.

   - **Example**:

     - Legislative bodies organized with \phi-based proportions allocate representation according to self-similar patterns, ensuring that both majority and minority groups have equitable influence.

2. **Balanced Leadership**:

   - Decision-making hierarchies designed around \phi-based principles promote a balance of power between central authorities and local communities.

   - **Example**:

     - A government structured with \phi-aligned tiers of authority ensures that each level has proportional influence, preventing both over-centralization and fragmentation.

3. **Harmonic Decision-Making**:

   - By integrating \phi-modulated processes into deliberation and voting systems, governance structures can achieve decisions that resonate with collective priorities.

   - **Prediction**:

- \phi-optimized voting mechanisms reduce polarization by emphasizing common ground and shared goals.

Applications of \phi in Governance

1. **Dynamic Policy Systems**:

- Policies can be structured to evolve proportionally with societal changes, maintaining harmony between governance and the needs of the population.

- **Example**:

  - A tax system based on \phi-aligned brackets ensures proportional contributions without overburdening any segment of the population.

2. **Crisis Management**:

- \phi-based frameworks can guide resource allocation during emergencies, ensuring balanced and efficient responses.

- **Example**:

  - A disaster relief program uses \phi-aligned logistics to distribute resources equitably, prioritizing the most affected regions.

3. **Intergovernmental Cooperation**:

- The principles of \phi can inspire proportional representation in global institutions, fostering collaboration on international challenges like climate change and public health.

- **Example**:

  - A United Nations voting system based on \phi-weighted contributions balances the influence of nations according to their population, economy, and environmental impact.

A Vision of Balanced Leadership

Governance aligned with \phi redefines leadership as a dynamic interplay between central authority and grassroots empowerment. These systems:

- **Distribute Power**: Proportional structures ensure fairness and inclusivity.

- **Enhance Adaptability**: Harmonized frameworks enable systems to evolve with changing societal needs.

- **Promote Unity**: Decision-making processes focus on shared priorities, reducing division and fostering collective progress.

The Future of Governance

Imagine a world where governance reflects the universal principles of harmony—where power is neither concentrated nor fragmented, and decisions resonate with the collective will. By integrating \phi-based principles into governance systems, we create structures that are not only more efficient but also deeply aligned with the natural balance of the cosmos. These systems pave the way for societies that thrive on cooperation, equity, and shared purpose.

Subsection 2: Economic Systems Inspired by \phi

The Challenges of Modern Economics

Economic systems shape the foundation of human societies, influencing resource distribution, innovation, and social equity. However, modern economies often exhibit patterns of imbalance:

1. **Wealth Concentration**: Resources disproportionately accumulate in the hands of a few, leaving vast inequalities in their wake.

2. **Environmental Degradation**: Economic growth often comes at the expense of ecological balance, exacerbating climate and resource crises.

3. **Volatility**: Financial systems are prone to booms and busts, creating cycles of instability that harm individuals and nations alike.

These challenges demand an economic paradigm that mirrors the self-regulating, sustainable patterns found in nature. The **Golden Ratio (\phi)**, with its inherent balance and proportionality, offers a transformative framework for reimagining how economies can operate in harmony with human and ecological needs.

The Role of \phi in Economic Systems

The principles of \phi provide a natural foundation for designing economic systems that are equitable, resilient, and sustainable.

1. **Proportional Wealth Distribution**:

- Economic structures inspired by \phi ensure that resources flow in a way that reflects the self-similar patterns found in nature.

- **Example**:

- A ϕ-aligned tax system allocates wealth contributions and redistributions in proportion to income growth, encouraging equity without stifling innovation.

2. **Sustainable Growth**:

- By modeling economic expansion on ϕ-based fractal growth patterns, systems can balance innovation with environmental preservation.

- **Example**:

  - An economy structured to grow within ϕ-aligned environmental limits avoids resource overexploitation, fostering long-term sustainability.

3. **Market Stability**:

- Financial systems designed with ϕ-inspired feedback loops can mitigate volatility by harmonizing supply, demand, and investment flows.

- **Prediction**:

  - ϕ-modulated market algorithms reduce boom-bust cycles by maintaining proportional adjustments in pricing and asset distribution.

Applications of ϕ-Based Economic Principles

1. **Resource Allocation**:

- ϕ-inspired frameworks prioritize resource distribution that balances societal needs with ecological limits.

- **Example**:

  - A national budgeting system uses ϕ-aligned allocation models to ensure proportional investment in healthcare, education, and infrastructure.

2. **Circular Economies**:

- Circular economic systems, which emphasize reuse and regeneration, align with the recursive and sustainable nature of ϕ.

- **Example**:

  - A ϕ-based circular economy maximizes resource efficiency by mirroring natural cycles, such as nutrient flows in ecosystems.

3. **Digital Economies**:

- As digital assets and cryptocurrencies rise, \phi-inspired tokenization systems can ensure proportional value creation and exchange.

  - **Example**:

    - A blockchain network designed with \phi-based protocols achieves equitable transaction fees and sustainable growth.

A Vision for Harmonious Economics

Economic systems inspired by \phi transcend the limitations of current paradigms, offering solutions that are both innovative and grounded in universal balance. These systems:

- **Promote Equity**: Wealth and resources flow proportionally, reducing inequality and fostering societal well-being.

- **Ensure Sustainability**: Growth models align with ecological principles, preserving resources for future generations.

- **Enhance Resilience**: Proportional feedback loops stabilize markets, preventing extreme volatility.

The Future of Economics

Imagine an economy where resources circulate like the flow of energy in a forest, balanced and self-sustaining. In this vision, businesses grow without depleting the environment, wealth is distributed without creating stagnation, and innovation thrives in harmony with human and planetary needs.

By integrating \phi-based principles into economic systems, we can create markets that reflect the timeless wisdom of the cosmos—a world where prosperity and harmony are not opposing forces but complementary facets of human progress.

Subsection 3: Communities and Urban Design with \phi

The Challenges of Modern Urbanization

Cities and communities are the physical manifestations of human civilization, but modern urban design often struggles to balance the needs of people, the environment, and technology:

1. **Overcrowding**: Urban centers face growing populations, leading to congestion and a strain on resources.

2. **Environmental Impact**: Expanding cities disrupt ecosystems, contributing to pollution and climate change.

3. **Disconnected Design**: Many urban layouts prioritize efficiency at the expense of community cohesion and aesthetic harmony.

To address these challenges, urban design must evolve to reflect the harmonious and self-sustaining patterns found in nature. The **Golden Ratio (\phi)**, with its universal principles of balance and proportion, offers a visionary framework for creating communities that thrive on beauty, sustainability, and interconnectedness.

The Role of \phi in Urban Design

By applying \phi-based principles, communities and cities can achieve layouts and systems that prioritize well-being, efficiency, and environmental harmony.

1. **Spiral and Fractal Layouts**:

- Urban layouts inspired by \phi-based spirals and fractals mimic natural growth patterns, optimizing space and enhancing connectivity.

- **Example**:

   - A city designed with \phi-spiral streets reduces congestion and improves accessibility by harmonizing the flow of people and vehicles.

2. **Green Spaces and Architecture**:

- Incorporating \phi-proportioned green spaces and buildings creates environments that promote mental health and aesthetic appeal.

- **Example**:

   - A community park arranged with \phi-aligned pathways and vegetation fosters tranquility while maximizing usable space.

3. **Sustainable Infrastructure**:

- Infrastructure designed around \phi-modulated proportions ensures efficient energy use and minimal environmental impact.

- **Example**:

   - A \phi-inspired water distribution network reduces waste and ensures equitable access to resources.

Applications of \phi-Based Community Design

1. **Smart Cities**:

• Integrating \phi-inspired layouts into smart city planning aligns digital technologies with human needs and natural rhythms.

• **Example**:

• A smart city grid using \phi-modulated traffic systems reduces congestion and pollution while improving mobility.

2. **Resilient Communities**:

• Communities designed with \phi-proportioned housing and resource distribution are better equipped to withstand economic and environmental challenges.

• **Example**:

• A coastal town uses \phi-aligned flood defenses and renewable energy systems to adapt to climate change.

3. **Global Collaboration**:

• Urban planners and architects worldwide can adopt \phi-based principles to create culturally unique yet universally harmonious designs.

• **Example**:

• A collaborative housing project incorporates \phi-inspired modular units, combining flexibility with aesthetic appeal.

A Vision of Living Harmony

Communities and cities designed with \phi are more than functional—they are vibrant ecosystems where people, nature, and technology coexist in balance. These designs:

• **Foster Connection**: \phi-aligned layouts promote social interaction and community cohesion.

• **Enhance Sustainability**: Proportional designs reduce waste and environmental impact, ensuring long-term viability.

• **Elevate Quality of Life**: Harmonious architecture and infrastructure create spaces that inspire creativity, relaxation, and well-being.

The Future of Urban Design

Imagine a city where every street, building, and park reflects the harmony of the natural world—a city where beauty and functionality are seamlessly intertwined. In this vision, communities are not just places to live but living systems that grow, adapt, and thrive in resonance with the universe.

By embedding \phi into urban design, we create environments that mirror the elegance of nature, paving the way for a future where humanity and the planet flourish together.

Conclusion: Societies Aligned with Universal Harmony

Society is the reflection of humanity's collective aspirations, challenges, and ingenuity. Yet, the systems we rely on often prioritize short-term gains over long-term sustainability, creating imbalances that strain both people and the planet. By integrating the principles of the **Golden Ratio (\phi)** and the **Etheric Phi Gravitation Formula (EPGF)** into governance, economics, and urban design, we can reimagine societal systems as living embodiments of balance, fairness, and interconnectedness.

Key Insights from Section 4

1. **Governance Aligned with \phi:**

   • Proportional representation, balanced leadership, and harmonic decision-making foster fairness and reduce polarization. These systems empower both individuals and communities while maintaining societal cohesion.

2. **Economic Systems Inspired by \phi:**

   • By aligning resource distribution and growth with \phi-based patterns, economies become more equitable, sustainable, and resilient. This approach balances innovation with environmental preservation, creating long-term stability.

3. **Communities and Urban Design:**

   • Cities and communities designed with \phi-inspired principles reflect the harmony of nature. From spiral layouts to sustainable infrastructure, these designs prioritize well-being, connectivity, and environmental stewardship.

A Vision for Societal Transformation

When societal systems mirror the harmony of the cosmos, they unlock new possibilities for human progress:

- **Equity and Inclusion**:

  - Proportional systems ensure that all voices are heard and resources are distributed fairly.

- **Sustainability**:

  - Designs that align with \phi minimize waste and environmental impact, fostering a regenerative relationship with the Earth.

- **Resilience and Innovation**:

  - By embracing natural patterns, societal systems become more adaptable to change and better equipped to navigate future challenges.

A Call to Action

To create a world where societal systems resonate with universal harmony, collective effort is essential:

1. **Policy and Leadership**:

   - Decision-makers must prioritize fairness, adaptability, and sustainability in governance and economic strategies.

2. **Collaboration Across Disciplines**:

   - Architects, urban planners, economists, and community leaders can work together to integrate \phi-based principles into their designs and policies.

3. **Empowering Communities**:

   - Grassroots movements can advocate for harmonic systems that reflect the needs and values of their populations.

Toward a Harmonious Future

Imagine a world where societies operate not through conflict and imbalance but through cooperation and shared purpose. By aligning our systems with the timeless wisdom of \phi, we create a future where humanity thrives in harmony with itself and the natural world. This is not a distant utopia—it is a tangible vision, ready to be realized through bold innovation and unwavering commitment to balance and equity.

With \phi as our guide, we step into a new era of societal transformation, where every decision, policy, and design reflects the elegant order of the universe.

Introduction: Personal Growth Aligned with Universal Principles

At the heart of every great transformation—whether societal, technological, or environmental—lies the individual. Personal growth is the foundation upon which human progress is built. Yet, in an age of relentless demands and distractions, the path to self-discovery, resilience, and fulfillment can feel elusive.

What if the key to personal growth lies not in reinventing ourselves but in rediscovering the harmony that already exists within us? By aligning our lives with the principles of the **Golden Ratio (**\phi**)**, we can tap into a natural rhythm that promotes balance, creativity, and resilience. The same proportionality and interconnectedness that shape galaxies, ecosystems, and art can guide us toward a deeper understanding of ourselves and our potential.

Personal Growth Through the Lens of \phi

1. **Balance and Proportion**:

   - Just as \phi structures nature's patterns, it can guide us in balancing work, relationships, and self-care.

2. **Harmonic Thinking**:

   - Embracing \phi's recursive patterns encourages self-reflection and continuous improvement.

3. **Resilience Through Connection**:

   - Aligning with the natural flow of \phi fosters a sense of interconnectedness, helping us navigate challenges with grace.

Exploring the Subsections

This section will delve into:

- How \phi-based principles can help individuals cultivate balance and harmony in their daily lives.

- Techniques for fostering resilience by aligning personal goals with natural patterns.

- The role of \phi in unlocking creativity and purpose, enabling individuals to thrive in an ever-changing world.

By applying the timeless wisdom of \phi to personal growth, we uncover a transformative path that aligns our inner lives with the universal harmony of the cosmos. Let us now explore how this ancient principle can illuminate the way forward.

Subsection 1: Cultivating Balance and Harmony Through \phi

The Challenge of Modern Life

In today's fast-paced world, the pursuit of balance often feels like an elusive goal. People struggle to juggle the demands of work, relationships, health, and personal fulfillment, leading to stress, burnout, and dissatisfaction. Yet, balance is not a static state—it is a dynamic rhythm, much like the natural cycles that govern the universe.

The **Golden Ratio (\phi)** offers a profound framework for achieving balance by mirroring the proportionality and harmony found in nature. Just as \phi governs the growth of trees, the spirals of galaxies, and the flow of rivers, it can guide us in structuring our lives in a way that promotes equilibrium and well-being.

Aligning Life with \phi

The principles of \phi can help us identify and maintain balance across the key dimensions of life.

1. **Proportional Time Management**:

   • Effective time management is less about rigid schedules and more about allocating time in a way that reflects our priorities and values. By dividing our day using \phi-based proportions, we can ensure that energy flows harmoniously between work, rest, and relationships.

   • **Example**:

      • A \phi-aligned schedule might allocate 62% of active hours to work or focused tasks, 38% to personal care, social activities, or creative pursuits, mirroring the natural balance found in \phi.

2. **Harmonic Relationships**:

   • Relationships thrive on balance—giving and receiving, listening and speaking, independence and connection. By aligning interactions with the principles of \phi, we can cultivate deeper and more fulfilling relationships.

   • **Example**:

- A mindful approach to communication involves listening for \phi-proportioned time during conversations, ensuring that both parties feel heard and valued.

3. **Holistic Well-Being**:

- Physical, mental, and emotional health are interconnected, and imbalance in one area often disrupts the others. Structuring self-care routines around \phi ensures that each aspect of well-being receives proportional attention.

- **Example**:

  - A fitness regimen based on \phi might allocate time proportionally between strength training (62%) and restorative practices like yoga or stretching (38%).

Techniques for Cultivating Balance with \phi

1. \phi-**Based Reflection**:

- Periodically evaluate the balance in your life using \phi-inspired questions:

  - Are your commitments aligned with your priorities?

  - Do your relationships reflect mutual harmony and respect?

- Use these reflections to adjust and recalibrate.

2. **Visualizing Harmony**:

- Create a visual representation of your time or energy distribution using a \phi-spiral diagram. This exercise provides a clear and intuitive view of where adjustments are needed.

3. **Setting Intentional Boundaries**:

- Proportional boundaries based on \phi help maintain harmony between external demands and personal needs.

- **Example**:

  - Designating a fixed portion of the day (e.g., 38%) as uninterrupted personal time fosters self-care and mental clarity.

The Benefits of a Balanced Life

By structuring life according to \phi, individuals experience:

- **Greater Clarity**: Harmonized priorities reduce stress and decision fatigue.

- **Improved Relationships**: Proportional attention fosters deeper connections with others.

- **Enhanced Well-Being**: A balanced approach to health and self-care creates lasting resilience and vitality.

Living in Harmony with the Universe

Balance is not an endpoint but a dynamic process of alignment. By integrating \phi-based principles into daily life, we not only achieve greater harmony but also align ourselves with the rhythms of the cosmos. This alignment fosters a profound sense of connection—to ourselves, to others, and to the larger world.

With \phi as a guide, balance becomes not just a goal but a natural and achievable state of being, one that empowers us to navigate life with grace and purpose.

Subsection 2: Building Resilience Through Universal Patterns

The Need for Resilience

Resilience is the ability to adapt, recover, and thrive in the face of adversity. In a world characterized by constant change and uncertainty, resilience is not just a desirable trait—it is essential for personal and collective survival. However, many people struggle to develop resilience due to fragmented thinking, emotional overwhelm, and a lack of connection to enduring principles.

The **Golden Ratio (\phi)** offers a framework for building resilience by aligning our responses to challenges with the patterns of the natural world. Just as nature demonstrates resilience through proportional growth and self-similarity, we too can learn to navigate difficulties by embracing the rhythms and structures of \phi.

How \phi Enhances Resilience

By mirroring the proportionality and harmony of \phi, individuals can cultivate resilience that is both adaptable and enduring.

1. **Self-Similar Growth**:

   - Resilience is not about returning to a previous state but about growing through adversity. The fractal nature of \phi demonstrates that growth occurs through repeated patterns at different scales.

- **Example:**

  - After experiencing a setback, focusing on small, \phi-proportioned improvements (e.g., 62% rebuilding core strengths, 38% exploring new opportunities) ensures sustainable growth.

2. **Balanced Recovery:**

- Recovery from stress or trauma requires a balance between action and rest. Structuring recovery periods using \phi-aligned cycles helps maintain this equilibrium.

- **Example:**

  - Allocating recovery time proportionally, such as 62% for physical rest and 38% for mental rejuvenation, fosters holistic healing.

3. **Harmonic Adaptation:**

- Resilience thrives on flexibility. By aligning responses to challenges with \phi-based timing and rhythm, individuals can adapt more effectively to changing circumstances.

- **Prediction:**

  - Aligning decision-making with \phi-modulated intervals (e.g., waiting 1.618 days to assess a major decision) reduces impulsivity and improves outcomes.

Techniques for Building Resilience with \phi

1. **Fractal Mindset Training:**

- Practice seeing challenges as opportunities for self-similar growth. Break large problems into smaller, proportional steps, addressing each with focus and intention.

- **Exercise:**

  - Identify a current challenge and divide your response efforts into \phi-proportioned steps, such as 62% action on immediate issues and 38% planning for long-term goals.

2. **Rhythmic Reflection:**

- Periodically reflect on personal progress and setbacks using \phi-aligned intervals. This rhythmic approach keeps growth in harmony with life's natural flow.

  - **Example**:

    - Set aside a 38% portion of your week for reflective practices like journaling or meditation to process challenges and identify patterns of growth.

3. **Anchoring in Nature**:

- Spend time observing natural systems that reflect \phi—from the spirals of shells to the branching of trees. These observations remind us of the resilience inherent in proportional and harmonious structures.

  - **Exercise**:

    - Take a walk in nature and identify patterns that mirror \phi, using them as metaphors for navigating your own challenges.

The Benefits of \phi-Inspired Resilience

By aligning personal growth and adaptation with \phi, individuals experience:

- **Greater Flexibility**: A proportional approach fosters adaptability without overextending resources.

- **Sustainable Growth**: Self-similar patterns ensure that progress is incremental and enduring.

- **Emotional Stability**: Harmonic rhythms reduce stress and promote mental clarity during challenges.

Thriving in Harmony with Adversity

Resilience is not about avoiding difficulty but about learning to navigate it with grace. By embracing the universal patterns of \phi, we align ourselves with a deeper, natural rhythm that supports us in overcoming life's obstacles. This alignment fosters not only recovery but also transformation, allowing us to emerge stronger and more harmonious with the world around us.

Through \phi, we discover that resilience is not an isolated skill but a continuous process of growth and renewal, a reflection of the universe's boundless capacity to adapt and thrive.

Subsection 3: Unlocking Creativity and Purpose with \phi

The Search for Creativity and Purpose

Creativity and purpose are deeply intertwined—each fuels the other, guiding individuals toward fulfilling lives. Yet, many struggle to access their creative potential or clarify their purpose, feeling disconnected from the deeper rhythms of life. In a world often driven by external expectations, the path to authentic self-expression and meaningful goals can feel obscured.

The **Golden Ratio (\phi)**, with its universal principles of harmony and proportionality, offers a framework for tapping into creativity and aligning with one's higher purpose. By understanding how \phi governs patterns in nature, art, and the cosmos, individuals can learn to express themselves and pursue goals in a way that feels natural, inspired, and enduring.

The Role of \phi in Creativity and Purpose

1. **Proportional Thinking**:

   • Creativity thrives on balance—between structure and spontaneity, effort and intuition. \phi provides a blueprint for navigating these dualities, fostering innovative and harmonious outcomes.

   • **Example**:

      • An artist might structure their work using \phi-proportioned sections, allowing for a balance of detailed focus (62%) and free exploration (38%).

2. **Self-Similar Expression**:

   • Purpose unfolds through recurring patterns of growth and self-discovery. Just as fractals repeat across scales, personal purpose reveals itself through recurring themes and passions.

   • **Exercise**:

      • Reflect on your life's defining moments and identify recurring patterns or motifs. Use these self-similar insights to guide future decisions and creative pursuits.

3. **Harmonic Flow**:

   • Aligning creative efforts with \phi-based timing and rhythm helps individuals enter a state of flow, where ideas and actions unfold effortlessly.

- **Prediction**:

  - \phi-aligned workflows could enhance productivity and creativity by synchronizing efforts with natural cycles.

Techniques for Unlocking Creativity and Purpose

1. \phi-**Guided Brainstorming**:

   - Use \phi-structured prompts to generate ideas. For example, divide brainstorming time into 62% focused ideation and 38% free association, mirroring \phi's proportions.

   - **Exercise**:

     - Begin by focusing on specific questions (e.g., "What inspires me?") before allowing open-ended exploration of ideas.

2. **Purpose Mapping**:

   - Create a visual map of your passions, values, and goals using a \phi-spiral diagram. This approach organizes your thoughts while revealing connections and priorities.

   - **Example**:

     - A purpose map might highlight how your creative skills align with community needs, guiding you toward meaningful projects.

3. **Harmonic Journaling**:

   - Dedicate a portion of your day to reflective writing, using \phi-aligned intervals to explore thoughts and intentions. This practice deepens self-awareness and clarifies purpose.

   - **Exercise**:

     - Spend 38% of your journaling session reflecting on past experiences and 62% envisioning future possibilities.

Applications Across Fields

1. **Art and Design**:

   - Artists and designers can use \phi-based principles to create works that resonate deeply with audiences, blending symmetry and emotion.

- **Example**:

    - A painter structures their composition using \phi-aligned focal points, creating balance and intrigue.

2. **Entrepreneurship**:

- Entrepreneurs can apply \phi to design purpose-driven ventures that align personal values with market needs.

    - **Example**:

        - A business plan structured around \phi balances innovation (62%) with practicality (38%), ensuring sustainable growth.

3. **Education and Personal Development**:

- Teachers and coaches can use \phi-based frameworks to inspire students and clients, fostering creativity and goal-setting.

    - **Example**:

        - A workshop uses \phi-inspired activities to help participants uncover and articulate their purpose.

The Benefits of \phi-Inspired Creativity and Purpose

By aligning creative expression and life goals with \phi, individuals experience:

- **Enhanced Innovation**: Proportional thinking sparks ideas that feel both novel and natural.

- **Clarity of Purpose**: Self-similar patterns reveal authentic priorities and aspirations.

- **Sustained Motivation**: Harmonic workflows maintain energy and focus over time.

A Life of Creative Harmony

Unlocking creativity and purpose is not about forcing inspiration—it is about aligning with the rhythms of the universe. By embracing the principles of \phi, individuals can access a wellspring of ideas and insights that feel both profound and effortless. In this state, creativity becomes a reflection of life's natural order, and purpose emerges as the guiding force that connects personal expression with the greater whole.

With \phi as our guide, we not only create but also discover meaning in the process, transforming our lives into works of art that resonate with the harmony of the cosmos.

Conclusion: Aligning Personal Growth with Universal Harmony

Personal growth is a journey that shapes not only individual lives but also the fabric of human progress. By aligning ourselves with the principles of the **Golden Ratio (\phi)**, we gain a profound framework for cultivating balance, resilience, and purpose. This alignment empowers us to navigate life's challenges with grace and transform them into opportunities for growth and creativity.

Key Insights from Section 5

1. **Balance and Harmony**:

   - \phi-based principles guide us in structuring our time, relationships, and well-being, fostering a dynamic and sustainable balance in life.

2. **Resilience Through Universal Patterns**:

   - By mirroring nature's fractal and proportional growth, we develop resilience that is both adaptable and enduring, allowing us to thrive in the face of adversity.

3. **Creativity and Purpose**:

   - Aligning our creative efforts and life goals with \phi unlocks new depths of innovation and clarity, connecting our personal expression to the greater rhythms of the universe.

The Promise of \phi-Inspired Growth

By embracing \phi-based principles, personal growth becomes more than a series of individual efforts—it becomes a harmonious dance with the universe itself. This alignment offers:

- **Clarity**: A deeper understanding of what matters most in life.

- **Flow**: The ability to move through challenges and opportunities with ease and intention.

- **Connection**: A sense of unity with the natural world and the larger human experience.

A Call to Action

To fully embrace the transformative power of \phi, individuals can:

1. **Reflect and Realign**:

   • Periodically assess your life through the lens of \phi, identifying areas where balance and harmony can be restored.

2. **Practice Harmonic Living**:

   • Incorporate \phi-based techniques into daily routines, from time management to creative pursuits.

3. **Inspire Others**:

   • Share the principles of \phi with your community, fostering a collective shift toward balance and interconnectedness.

A Life in Resonance

Imagine a life where each decision, relationship, and creative endeavor resonates with the timeless rhythms of the cosmos. In this vision, personal growth is not a solitary pursuit but a shared journey that connects us to one another and to the natural world. By aligning with \phi, we unlock our potential to live with purpose, resilience, and harmony, transforming our lives into reflections of the universe's infinite beauty.

With \phi as our guide, personal growth becomes a path not just to self-improvement but to universal connection—a journey that celebrates the profound order and creativity of existence itself.

Introduction: A Unified Approach to Knowledge and Innovation

Human progress has always been driven by the synthesis of ideas across disciplines. From the Renaissance to the Information Age, the greatest breakthroughs have emerged when diverse fields—art, science, philosophy, and technology—come together to create something greater than the sum of their parts. Yet, in our increasingly specialized world, this integration often feels fragmented, leaving us searching for connections that can bridge the gaps.

The **Golden Ratio (\phi)** provides a universal framework for interdisciplinary thinking, offering a lens through which we can unify knowledge and innovation. Whether in architecture, medicine, artificial intelligence, or cultural development,

\phi's principles of balance, proportionality, and harmony serve as a guide for aligning diverse systems and ideas with the natural order.

This section explores how \phi can inspire integration across disciplines, creating solutions that are not only innovative but also deeply harmonious with the rhythms of life and the cosmos.

The Power of Interdisciplinary Integration

1. **Bridging the Gaps**:

   - \phi serves as a common denominator across fields, providing a shared language of proportionality and balance.

2. **Enhancing Creativity**:

   - By applying \phi to interdisciplinary projects, we unlock novel insights and solutions that transcend traditional boundaries.

3. **Fostering Sustainability**:

   - Integrated systems designed with \phi prioritize harmony and resilience, ensuring long-term viability.

Exploring the Subsections

This section will delve into:

- The role of \phi in unifying scientific and artistic disciplines, fostering breakthroughs in design and technology.

- Applications of \phi in medicine and healthcare, where proportionality and balance enhance patient outcomes.

- How \phi can inspire cultural and societal systems that reflect a unified vision of human progress.

By embracing \phi as a unifying principle, we open the door to a new era of innovation—one where knowledge and creativity flow seamlessly across disciplines, guided by the harmony of the universe.

Subsection 1: \phi as a Bridge Between Art and Science

The Historical Divide Between Art and Science

Art and science are often seen as separate realms: one driven by creativity and intuition, the other by logic and experimentation. Yet, history reveals that the greatest innovations arise when these two disciplines converge. From Leonardo da Vinci's visionary inventions to the elegant simplicity of Einstein's equations, the interplay between art and science has shaped humanity's understanding of the world.

The **Golden Ratio (**ϕ**)** serves as a natural bridge between these disciplines. Found in the proportions of masterpieces and the structures of natural phenomena, ϕ unites the creative and analytical, offering a shared framework for exploration and discovery.

The Role of ϕ **in Unifying Art and Science**

1. **Proportional Beauty**:

- ϕ is celebrated in art for its ability to create aesthetically pleasing compositions, from the Parthenon to the paintings of the Renaissance.

- In science, ϕ governs the proportions of galaxies, DNA structures, and other natural systems.

- **Example**:

    - A modern architect designs a building using ϕ-aligned proportions to harmonize its visual appeal with structural integrity, blending artistic expression with engineering precision.

2. **Harmonic Design**:

- Both art and science rely on patterns and structures to achieve their goals. ϕ's recursive and self-similar properties provide a framework for creating designs that are both functional and beautiful.

- **Example**:

    - A biomedical engineer develops a prosthetic limb inspired by the ϕ-based proportions found in human anatomy, enhancing both functionality and form.

3. **Inspiration from Nature**:

- Nature's use of ϕ demonstrates how beauty and efficiency coexist, inspiring both artists and scientists to emulate these patterns.

- **Example**:

- A product designer creates a solar panel array modeled after \phi-aligned sunflower spirals, maximizing energy efficiency while celebrating natural aesthetics.

Applications Across Disciplines

1. **Innovative Architecture**:

- Architects and engineers can use \phi-based designs to create structures that are resilient, sustainable, and visually harmonious.

- **Example**:

- A city planner incorporates \phi-spiral layouts into urban development, creating spaces that balance functionality and beauty.

2. **Biomedical Advancements**:

- The proportionality of \phi can guide innovations in medical imaging, device design, and treatment strategies.

- **Example**:

- A surgeon uses \phi-based mapping techniques to improve the precision of reconstructive procedures, aligning with the body's natural proportions.

3. **Digital Creativity**:

- In fields like video game design and virtual reality, \phi-inspired geometries create immersive and visually stunning experiences.

- **Example**:

- A video game designer incorporates \phi-aligned landscapes, enhancing player engagement through naturalistic and harmonious environments.

A New Era of Collaboration

By embracing \phi as a unifying principle, art and science can transcend traditional boundaries:

- **Enhanced Creativity**: The integration of \phi sparks innovation by blending intuition with analysis.

- **Sustainable Solutions**: Proportional designs prioritize harmony, reducing waste and inefficiency.

- **Universal Appeal**: Projects inspired by \phi resonate with people on a deeper level, connecting aesthetics and function.

The Promise of Integration

In a world increasingly defined by specialization, \phi offers a reminder that the greatest progress occurs at the intersection of disciplines. By bridging art and science through the principles of harmony and proportion, we unlock a new realm of possibility—one where creativity and logic work together to create beauty, solve problems, and inspire humanity.

With \phi as our guide, we can create a future where art and science are no longer separate endeavors but unified expressions of the same universal truth.

Subsection 2: \phi in Medicine and Healthcare

The Complexity of Modern Medicine

Medicine has always sought to balance precision with compassion, science with humanity. However, modern healthcare systems face significant challenges:

1. **Personalization**: Standardized treatments often fail to account for individual differences in biology and lifestyle.

2. **Integration**: Fragmented systems of care lead to inefficiencies and poor patient outcomes.

3. **Sustainability**: The environmental and financial costs of healthcare continue to rise, demanding innovative approaches.

The **Golden Ratio (**\phi**)** offers a framework for addressing these challenges by aligning medical practices and healthcare systems with the natural principles of harmony and proportionality. From patient care to systemic design, \phi provides a unifying approach that fosters efficiency, sustainability, and holistic well-being.

The Role of \phi in Medicine

1. **Proportional Anatomical Understanding**:

   - Human anatomy reflects \phi in countless ways, from the proportions of the face to the structure of DNA. Leveraging these natural patterns enhances medical precision.

   - **Example**:

- A reconstructive surgeon uses \phi-based templates to guide procedures, ensuring outcomes that are both functional and aesthetically harmonious.

2. **Harmonic Diagnostics**:

- Diagnostic systems designed around \phi-based principles can improve accuracy by aligning with the body's natural rhythms and proportions.

- **Example**:

- A cardiologist uses \phi-aligned algorithms to analyze heart rhythms, identifying anomalies with greater precision.

3. **Sustainable Healthcare Systems**:

- By incorporating \phi-based designs into hospital layouts, resource allocation, and care delivery, healthcare systems can reduce waste and improve patient experiences.

- **Example**:

- A hospital redesigns its floor plan using \phi-spirals to optimize patient flow and reduce stress for both staff and patients.

Applications in Modern Medicine

1. **Personalized Medicine**:

- \phi-inspired frameworks can guide the development of treatments tailored to individual patients, reflecting their unique biological proportions.

- **Example**:

- A gene therapy approach uses \phi-aligned models to predict the most effective interventions based on a patient's genetic structure.

2. **Advanced Medical Imaging**:

- Medical imaging systems, such as MRIs and CT scans, can use \phi-modulated algorithms to enhance image clarity and diagnostic accuracy.

- **Example**:

- An MRI machine incorporates \phi-based signal processing to improve the resolution of complex tissue structures.

3. **Regenerative Medicine**:

- Tissue engineering and regenerative therapies benefit from \phi-aligned scaffolding, which mirrors the body's natural growth patterns.

- **Example**:

  - A bioengineer designs a \phi-inspired scaffold for tissue regeneration, improving cell growth and structural stability.

The Benefits of \phi-Inspired Healthcare

1. **Enhanced Precision**:

   - Proportional tools and techniques lead to more accurate diagnostics and treatments.

2. **Holistic Care**:

   - \phi-aligned systems prioritize the integration of mind, body, and environment, fostering overall well-being.

3. **Sustainability**:

   - By aligning with natural principles, healthcare systems become more resource-efficient and environmentally friendly.

A New Era of Medicine

Imagine a healthcare system where every tool, treatment, and space reflects the harmony of \phi. In this vision:

- Patients receive care that resonates with their unique needs and natural rhythms.

- Healthcare professionals work within systems designed to reduce stress and enhance efficiency.

- Medical innovation draws inspiration from the elegance of nature, creating solutions that are both cutting-edge and deeply human.

By integrating \phi into medicine, we not only improve patient outcomes but also redefine healthcare as a practice rooted in harmony, sustainability, and universal connection.

Subsection 3: \phi-Inspired Cultural and Societal Development

The Role of Culture in Shaping Society

Culture is the soul of society. It influences how we think, behave, and connect with one another. Yet, modern cultural systems often reflect fragmentation, imbalance, and unsustainable practices:

1. **Loss of Identity**: Globalization, while fostering connectivity, can erode local traditions and values.

2. **Imbalance in Representation**: Cultural narratives often favor certain groups, leaving others marginalized.

3. **Unsustainable Practices**: Consumer-driven culture prioritizes short-term gratification over long-term well-being.

The **Golden Ratio (**ϕ**)**, as a symbol of harmony and proportionality, offers a pathway to reimagine cultural systems that nurture identity, inclusivity, and sustainability. By aligning cultural development with ϕ's principles, we can foster societies that thrive on balance, creativity, and shared purpose.

The Role of ϕ in Cultural Development

1. **Proportional Representation**:

- Cultural systems inspired by ϕ prioritize inclusivity, ensuring that diverse voices are represented in proportion to their significance within the community.

- **Example**:

  - A national curriculum designed using ϕ-based principles allocates equal focus to major historical events and marginalized perspectives, creating a balanced narrative.

2. **Harmonic Creativity**:

- Art, music, and literature aligned with ϕ resonate deeply with human emotions, fostering unity and shared experiences.

- **Example**:

  - A symphony composed with ϕ-proportioned intervals evokes a sense of universal harmony, uniting audiences across cultures.

3. **Sustainable Practices**:

- Cultural systems that mirror \phi embrace practices that balance consumption with preservation, fostering long-term sustainability.

  - **Example**:

    - A community art project uses recycled materials and \phi-aligned designs to create installations that celebrate local heritage.

Applications in Cultural and Societal Systems

1. **Education and Storytelling**:

   - Curricula and storytelling frameworks inspired by \phi can balance tradition and innovation, fostering a deeper understanding of human history and potential.

     - **Example**:

       - A digital storytelling platform structures its narratives using \phi-aligned arcs, engaging audiences with proportional pacing and themes.

2. **Cultural Festivals and Events**:

   - Festivals designed with \phi-based layouts and schedules enhance the flow of activities, creating immersive and meaningful experiences.

     - **Example**:

       - A cultural festival uses a \phi-spiral layout to guide visitors through exhibits, performances, and workshops, promoting exploration and connection.

3. **Urban and Rural Integration**:

   - Societal systems inspired by \phi harmonize urban development with rural preservation, ensuring that both thrive in proportion.

     - **Example**:

       - A regional development plan allocates resources based on \phi-aligned population and geographic ratios, fostering balanced growth.

The Benefits of \phi-Inspired Cultural Systems

1. **Strengthened Identity**:

- Proportional representation ensures that all cultural voices are valued, fostering a sense of belonging.

2. **Unified Communities**:

- Harmonic creativity bridges divides, uniting people through shared experiences and values.

3. **Sustainable Growth**:

- Practices aligned with \phi prioritize preservation and renewal, ensuring that cultural heritage thrives for generations.

A Vision for Cultural Harmony

Imagine a society where cultural systems reflect the elegance and balance of nature—where traditions are preserved without stifling innovation, and creativity flourishes without sacrificing sustainability. In this vision:

- Communities celebrate their unique identities while embracing shared human values.

- Artistic expression inspires connection and understanding across borders.

- Societal growth mirrors the rhythms of \phi, balancing progress with preservation.

By integrating \phi into cultural and societal development, we create systems that resonate deeply with the human spirit, fostering a sense of unity and purpose that transcends time and place.

Conclusion: A Unified Framework for Knowledge and Progress

Integration is the cornerstone of human advancement. The ability to connect ideas across disciplines has led to the most profound breakthroughs in history, from the Renaissance to the modern technological era. By embracing the **Golden Ratio (\phi)** as a unifying principle, we unlock a new era of interdisciplinary innovation—one that aligns our creations with the harmony and balance of the universe.

Key Insights from Section 6

1. **Art and Science**:

- \phi-inspired frameworks bridge the gap between artistic intuition and scientific precision, creating innovations that are both functional and beautiful.

2. **Medicine and Healthcare**:

   • Proportional and harmonic designs enhance patient care, system efficiency, and sustainability, redefining healthcare as a practice of holistic well-being.

3. **Cultural and Societal Development**:

   • \phi-aligned cultural systems promote inclusivity, creativity, and sustainability, fostering societies that thrive on shared purpose and harmony.

The Benefits of Integration Through \phi

1. **Enhanced Creativity**:

   • By blending diverse perspectives and fields, \phi-based systems unlock innovative solutions that transcend traditional boundaries.

2. **Sustainability and Resilience**:

   • Designs and practices inspired by \phi prioritize balance, ensuring long-term viability and adaptability.

3. **Universal Connection**:

   • \phi serves as a common language of harmony, fostering collaboration and understanding across disciplines, cultures, and nations.

A Call to Action

The potential of \phi to inspire interdisciplinary integration requires intentional effort and collaboration:

1. **Fostering Collaboration**:

   • Encourage partnerships between artists, scientists, engineers, and cultural leaders to explore \phi-inspired approaches.

2. **Investing in Research**:

   • Support studies and pilot projects that apply \phi to real-world challenges, from urban design to healthcare.

3. **Celebrating Shared Values**:

   • Promote \phi-aligned systems and creations that emphasize universal harmony, beauty, and balance.

A Vision for Unified Progress

Imagine a world where every field of knowledge and practice resonates with the timeless rhythms of the universe—a world where art and science, medicine and culture, innovation and tradition come together to create a harmonious future. In this vision, \phi is not just a mathematical constant but a guiding force that inspires humanity to transcend fragmentation and embrace unity.

By integrating \phi across disciplines, we create systems that are not only innovative but also deeply aligned with the natural order of the cosmos. This alignment fosters a future where human progress and universal harmony go hand in hand, transforming our world into a reflection of the beauty and balance we find in nature.

Introduction: Ethical Leadership Rooted in Universal Principles

Leadership is the driving force behind human progress. It shapes the decisions that guide societies, businesses, and nations, influencing the lives of millions. However, in a world grappling with challenges like inequality, climate change, and political polarization, leadership often falls short of the balance and foresight required to address these issues effectively.

What if ethical leadership could draw inspiration from the natural harmony of the universe? The **Golden Ratio (\phi)** provides a timeless blueprint for aligning leadership with universal principles of balance, proportionality, and interconnectedness. By integrating \phi into leadership philosophies and practices, we can cultivate leaders who inspire trust, foster collaboration, and create lasting positive impact.

This section explores how \phi-aligned leadership can transform global systems, from politics and business to education and environmental stewardship, promoting a future rooted in equity, sustainability, and harmony.

Leadership Through the Lens of \phi

1. **Proportional Decision-Making**:

   • Leaders inspired by \phi allocate resources and responsibilities in ways that balance immediate needs with long-term goals.

2. **Harmonic Relationships**:

   • \phi-aligned leaders prioritize connections and collaboration, fostering trust and mutual respect.

3. **Sustainable Vision**:

- Leadership guided by \phi emphasizes holistic, long-term solutions that benefit people and the planet.

Exploring the Subsections

This section will delve into:

- How \phi-based principles inspire ethical decision-making and governance.

- The role of \phi in fostering trust and collaboration in business and politics.

- Applications of \phi in addressing global challenges like climate change and social inequality.

By embracing \phi as a guiding principle, leadership becomes more than a position of authority—it becomes a practice of alignment with the natural order, empowering leaders to create meaningful and lasting change.

Subsection 1: Ethical Decision-Making and Governance Inspired by \phi

The Challenges of Modern Governance

Governance is central to shaping societies and addressing global challenges. Yet, ethical decision-making often falters under the pressures of short-term gains, political polarization, and competing interests. These issues lead to:

1. **Imbalanced Resource Allocation**: Policies often prioritize the needs of a few over the well-being of the majority.

2. **Lack of Long-Term Vision**: Decision-making frequently sacrifices future sustainability for immediate results.

3. **Erosion of Trust**: Corruption, inequality, and inefficiency undermine public confidence in governance.

The **Golden Ratio (\phi)**, with its principles of harmony, proportionality, and balance, offers a foundation for reimagining ethical governance. By aligning decisions with \phi, leaders can create systems that prioritize equity, sustainability, and collective well-being.

The Role of \phi in Ethical Governance

1. **Proportional Policy Design**:

- Policies inspired by \phi balance the needs of different sectors and demographics, ensuring fair resource distribution.

  - **Example**:

    - A national budget allocates funds using \phi-proportioned ratios: 62% for essential services like healthcare and education, and 38% for infrastructure and innovation.

2. **Balanced Decision-Making**:

- \phi encourages leaders to weigh short-term needs against long-term impacts, creating decisions that stand the test of time.

  - **Example**:

    - Urban planning policies prioritize renewable energy adoption (62%) while addressing immediate energy demands (38%), fostering sustainability.

3. **Harmonic Representation**:

- Governance systems designed with \phi ensure proportional representation, fostering inclusivity and reducing conflict.

  - **Prediction**:

    - Electoral reforms based on \phi-aligned voting districts could reduce political polarization by emphasizing balanced representation.

Applications in Ethical Leadership

1. **Transparent Decision-Making**:

- Leaders can use \phi-based frameworks to communicate decisions transparently, building trust with stakeholders.

  - **Example**:

    - A government uses \phi-inspired visual models to explain policy impacts, ensuring clarity and accountability.

2. **Global Governance**:

- International organizations can adopt \phi-aligned frameworks to address global challenges like climate change and economic inequality.

- **Example**:

  - A global climate agreement divides responsibilities among nations based on \phi-proportioned metrics of emissions, resources, and capacity.

3. **Community-Led Initiatives**:

- Local governments can design participatory governance systems that mirror \phi, ensuring proportional input from diverse voices.

- **Example**:

  - A city council allocates 62% of decision-making power to elected representatives and 38% to community-driven initiatives, fostering collaboration.

The Benefits of \phi-Inspired Governance

1. **Equity and Fairness**:

- Proportional policies ensure that resources and opportunities are distributed equitably across populations.

2. **Sustainability**:

- Balanced decision-making supports long-term resilience and environmental stewardship.

3. **Trust and Transparency**:

- \phi-aligned systems foster trust by emphasizing clarity, accountability, and inclusivity.

A Vision for Ethical Leadership

Imagine a world where governance reflects the harmony of nature—where decisions are made not for short-term gains but for the lasting benefit of humanity and the planet. In this vision:

- Policies align with universal principles of fairness and sustainability.

- Leaders prioritize the well-being of future generations alongside present needs.

- Communities feel empowered, connected, and represented in the systems that govern them.

By integrating \phi into ethical decision-making, leaders can create governance structures that resonate with the natural order, fostering a future where equity, trust, and harmony prevail.

Subsection 2: Building Trust and Collaboration Through \phi

The Importance of Trust in Leadership

Trust is the cornerstone of effective leadership and collaboration. Without it, even the most well-intentioned initiatives can falter. Yet, in today's world, trust in leaders, institutions, and systems is at an all-time low due to:

1. **Broken Promises**: Inconsistencies between words and actions erode confidence.

2. **Lack of Transparency**: Decisions made behind closed doors breed suspicion and resistance.

3. **Polarization**: Divisive rhetoric and practices undermine collective action.

The **Golden Ratio (\phi)**, with its principles of harmony and proportionality, offers a framework for fostering trust and collaboration. By aligning actions, communications, and relationships with \phi, leaders can create environments where trust flourishes, enabling meaningful partnerships and collective progress.

The Role of \phi in Building Trust

1. **Proportional Communication**:

   - Trust thrives on balanced communication. \phi-aligned interactions ensure that listening and speaking occur in harmony, fostering mutual respect and understanding.

   - **Example**:

     - A leader dedicates 62% of a meeting to listening to team members' input and 38% to sharing their vision, creating a proportional dialogue.

2. **Harmonic Collaboration**:

   - \phi inspires collaboration frameworks that balance individual contributions with collective goals, ensuring that every voice is valued.

   - **Example**:

- A project team allocates tasks using \phi-proportioned criteria, balancing expertise (62%) with opportunities for growth (38%).

3. **Transparent Actions**:

- Leaders guided by \phi prioritize transparency and consistency, aligning their actions with promises and principles.

- **Example**:

    - A government uses \phi-structured progress reports to provide clear updates on policy implementation, fostering public trust.

Applications in Trust and Collaboration

1. **Business Leadership**:

- Corporate leaders can use \phi-based frameworks to build trust with employees, stakeholders, and customers.

- **Example**:

    - A company structures its employee engagement strategy with \phi-aligned investments: 62% in professional development and 38% in wellness initiatives.

2. **International Cooperation**:

- Diplomatic efforts can benefit from \phi-inspired proportionality, balancing national interests with global priorities.

- **Example**:

    - A climate agreement uses \phi-aligned contributions, ensuring that resource-rich nations provide proportional support to smaller, vulnerable countries.

3. **Community Initiatives**:

- Local leaders can apply \phi-based principles to create collaborative spaces that empower diverse voices and ideas.

- **Example**:

    - A city council uses \phi-proportioned allocations to fund community-led projects, balancing innovation (62%) with cultural preservation (38%).

The Benefits of \phi-Inspired Collaboration

1. **Strengthened Relationships**:

   • Proportional communication fosters deeper connections and mutual understanding.

2. **Increased Transparency**:

   • Clear, consistent actions aligned with \phi build credibility and trust.

3. **Empowered Teams**:

   • Balanced collaboration ensures that individual strengths are valued while collective goals are achieved.

A Vision for Collaborative Leadership

Imagine a world where leaders prioritize trust and collaboration as the foundation of progress. In this vision:

   • Teams work in harmony, balancing diverse perspectives with shared goals.

   • Decisions are made transparently, inspiring confidence and unity.

   • Communities and organizations thrive on mutual respect and proportional contributions.

By integrating \phi into leadership practices, we create environments where trust and collaboration are not just ideals but everyday realities. These environments enable individuals, teams, and nations to achieve their highest potential—together.

Subsection 3: Addressing Global Challenges Through \phi

The Scale of Global Challenges

Humanity faces unprecedented global challenges, from climate change and resource scarcity to social inequality and geopolitical conflict. These challenges demand solutions that are not only innovative but also inclusive, sustainable, and scalable. However, existing approaches often:

1. **Prioritize Short-Term Gains**: Immediate results are pursued at the expense of long-term stability.

2. **Lack Holistic Perspectives**: Siloed efforts fail to account for the interconnected nature of global issues.

3. **Exacerbate Inequalities**: Many initiatives disproportionately benefit wealthy nations and individuals, leaving vulnerable populations behind.

The **Golden Ratio (**ϕ**)**, with its principles of balance, harmony, and proportionality, provides a framework for addressing these challenges in ways that align with the natural order of the universe. By integrating ϕ-based principles into global strategies, we can create solutions that are both effective and equitable.

The Role of ϕ in Addressing Global Challenges

1. **Proportional Resource Allocation**:

   • Global efforts often fail due to uneven distribution of resources. ϕ-aligned allocation ensures proportional support for diverse regions and populations.

   • **Example**:

      •   An international aid program distributes funds using ϕ-proportioned criteria, allocating 62% to high-need areas and 38% to preventative measures.

2. **Harmonic Systems Design**:

   • Complex challenges require systems that balance efficiency with adaptability. ϕ-inspired designs create harmonized frameworks that account for multiple variables.

   • **Example**:

      •   A global water management initiative incorporates ϕ-based flow patterns to optimize usage and reduce waste across regions.

3. **Long-Term Sustainability**:

   • ϕ emphasizes proportional growth and renewal, guiding efforts that prioritize long-term impact over immediate results.

   • **Example**:

      •   A reforestation project uses ϕ-aligned planting patterns to enhance biodiversity and ensure ecological balance.

Applications in Global Solutions

1. **Climate Action**:

- \phi-based frameworks can guide climate initiatives that balance mitigation, adaptation, and restoration efforts.

  - **Example**:

    - A carbon reduction strategy allocates 62% of resources to renewable energy and 38% to reforestation, ensuring a balanced approach.

2. **Economic Equity**:

- Global economic systems can adopt \phi-aligned models to reduce inequality and promote sustainable development.

  - **Example**:

    - A global tax policy uses \phi-proportioned tiers to ensure fair contributions from wealthier nations while supporting developing economies.

3. **Conflict Resolution**:

- Diplomatic frameworks inspired by \phi foster proportional representation and collaboration, reducing the likelihood of conflict.

  - **Example**:

    - A peace agreement uses \phi-aligned negotiation phases, balancing immediate ceasefire measures (62%) with long-term reconciliation efforts (38%).

The Benefits of \phi-Inspired Global Strategies

1. **Equity and Inclusivity**:

- Proportional approaches ensure that all voices and needs are considered in global efforts.

2. **Sustainability**:

- \phi-aligned strategies prioritize long-term stability and resource renewal.

3. **Collaboration**:

- By emphasizing harmony, \phi fosters global partnerships that transcend individual interests.

A Vision for Global Harmony

Imagine a world where global challenges are addressed not through competition and division but through collaboration and shared purpose. In this vision:

- Nations work together, guided by proportional contributions and equitable outcomes.

- Environmental restoration aligns with the natural rhythms of the planet, fostering resilience and renewal.

- Social systems prioritize harmony, ensuring that no one is left behind.

By integrating \phi into global strategies, humanity can move beyond short-term fixes to create solutions that resonate with the balance and beauty of the cosmos. These solutions empower us to address today's challenges while building a sustainable, equitable future for generations to come.

Conclusion: Leading with Harmony and Universal Principles

Leadership has always been a powerful force for shaping human destiny. In a world fraught with complex challenges, the need for ethical, visionary leadership has never been greater. By aligning leadership with the principles of the **Golden Ratio (\phi)**, we unlock the potential to address global challenges with balance, equity, and long-term vision.

Key Insights from Section 7

1. **Ethical Decision-Making**:

   - \phi-aligned governance fosters proportional resource allocation, balanced policies, and decisions that prioritize both immediate needs and future sustainability.

2. **Trust and Collaboration**:

   - Leaders inspired by \phi emphasize transparency, proportional communication, and collaborative frameworks, building trust and empowering collective progress.

3. **Global Solutions**:

   - Addressing challenges like climate change, inequality, and resource scarcity through \phi-based strategies ensures that solutions are inclusive, sustainable, and scalable.

The Benefits of \phi-Inspired Leadership

1. **Equity and Fairness**:

   • Proportional approaches ensure that all voices and needs are considered, fostering inclusive systems.

2. **Sustainability**:

   • Long-term thinking and balanced resource use create enduring solutions for global challenges.

3. **Unity and Collaboration**:

   • By aligning with universal principles, \phi-inspired leadership transcends divisions, promoting collective action.

A Call to Action

To fully realize the potential of \phi in leadership and global impact, we must:

1. **Embrace Proportional Thinking**:

   • Leaders across all sectors should integrate \phi-aligned principles into decision-making, communication, and strategy.

2. **Foster Interdisciplinary Collaboration**:

   • Global challenges require solutions that combine diverse perspectives and expertise, unified by the harmony of \phi.

3. **Lead by Example**:

   • Ethical leadership inspired by \phi starts with individual actions that reflect integrity, transparency, and inclusivity.

A Vision for Global Harmony

Imagine a world where leaders guide societies not through domination or division but through collaboration and shared purpose. In this vision:

• Governance is fair and transparent, prioritizing equity and sustainability.

• Communities and nations work together to address challenges, inspired by the balance and harmony of \phi.

• Global systems evolve in alignment with the natural rhythms of the universe, fostering a future where humanity thrives in harmony with itself and the planet.

By integrating \phi into ethical leadership, we create a new paradigm—one where human progress and universal harmony are inseparable. This is the future that awaits us when we lead with the principles of balance, proportionality, and interconnectedness.

Introduction: A Vision of Harmony and Progress

Humanity stands at a crossroads. The decisions we make today will shape the future of our planet and the generations to come. As we confront challenges like climate change, technological disruption, and social inequality, the need for a unifying framework has never been more urgent. How can we create a future that balances progress with sustainability, individuality with collective well-being, and innovation with timeless wisdom?

The **Golden Ratio (\phi)** offers a guiding principle for envisioning and building this future. Rooted in the universal patterns that govern nature, \phi provides a blueprint for creating systems that are harmonious, resilient, and adaptive. By aligning humanity's trajectory with \phi, we can transcend fragmentation and forge a future of unity, balance, and boundless potential.

This section explores how \phi can inspire humanity's evolution, guiding advancements in technology, society, and philosophy toward a future that reflects the harmony of the cosmos.

The Role of \phi in Humanity's Future

1. **Balancing Innovation and Ethics**:

   • \phi provides a framework for advancing technology and society in ways that respect both human values and environmental limits.

2. **Fostering Unity**:

   • By aligning global systems with \phi, we can bridge divides and create shared purpose across cultures and nations.

3. **Sustainability and Resilience**:

   • \phi-inspired systems prioritize long-term stability, ensuring that progress does not come at the cost of future generations.

Exploring the Subsections

This section will delve into:

- The role of \phi in shaping sustainable technological advancements and innovations.

- How \phi-aligned philosophies can guide humanity's collective growth.

- Applications of \phi in creating a unified vision for the future, grounded in harmony and balance.

Through the lens of \phi, humanity's future becomes a tapestry woven with the threads of progress, unity, and sustainability—a future where innovation and wisdom converge to create a harmonious world.

Subsection 1: Sustainable Innovation Guided by \phi

The Double-Edged Sword of Innovation

Innovation has been the driving force behind humanity's greatest achievements, from the advent of agriculture to the rise of artificial intelligence. Yet, unchecked innovation often leads to unintended consequences:

1. **Environmental Degradation**: Technological progress frequently prioritizes growth over sustainability, resulting in resource depletion and pollution.

2. **Social Disparities**: New technologies can exacerbate inequality, benefiting some while leaving others behind.

3. **Ethical Oversights**: Rapid advancements often outpace the ethical frameworks needed to guide their development.

The **Golden Ratio (\phi)**, with its principles of balance, proportionality, and harmony, offers a framework for fostering innovation that is both forward-thinking and sustainable. By aligning technological progress with \phi, we can ensure that advancements serve humanity and the planet in equal measure.

The Role of \phi in Sustainable Innovation

1. **Proportional Resource Allocation**:

   - Innovation guided by \phi ensures that resources are allocated in a balanced and efficient manner, prioritizing both short-term goals and long-term sustainability.

   - **Example**:

- A renewable energy initiative allocates 62% of resources to solar power development and 38% to energy storage solutions, creating a balanced system.

2. **Harmonic Design**:

- Technologies inspired by \phi integrate seamlessly with natural systems, minimizing disruption and maximizing efficiency.

- **Example**:

    - An urban vertical farm uses \phi-aligned layouts to optimize sunlight exposure and water distribution, reducing waste and enhancing productivity.

3. **Ethical Development**:

- By incorporating \phi-based principles into the design and deployment of new technologies, innovators can ensure that their creations align with universal values of equity and harmony.

- **Prediction**:

    - \phi-guided AI algorithms could balance efficiency with fairness, reducing bias and promoting inclusivity.

Applications in Sustainable Innovation

1. **Renewable Energy**:

- \phi-based frameworks can guide the development of energy systems that are efficient, scalable, and environmentally friendly.

- **Example**:

    - A wind turbine design uses \phi-spiral configurations to capture wind energy more effectively while reducing noise pollution.

2. **Circular Economies**:

- Innovation inspired by \phi can create systems that prioritize reuse, regeneration, and minimal waste.

- **Example**:

    - A manufacturing process uses \phi-aligned material flows to maximize efficiency and minimize resource waste.

3. **Artificial Intelligence and Robotics**:

- Technologies that mirror the proportional and adaptive nature of \phi can integrate more harmoniously into human lives and environments.

  - **Example**:

    - A robot designed with \phi-proportioned limbs and movements achieves greater efficiency and aesthetic appeal, enhancing user interactions.

The Benefits of \phi-Inspired Innovation

1. **Sustainability**:

   - Proportional designs reduce waste and resource overuse, ensuring that innovation supports long-term ecological balance.

2. **Equity and Accessibility**:

   - \phi-aligned systems prioritize inclusivity, making technological advancements accessible to diverse populations.

3. **Resilience**:

   - Innovations guided by \phi adapt more readily to changing circumstances, ensuring their relevance and effectiveness over time.

A Vision for Sustainable Progress

Imagine a world where every innovation—whether in energy, transportation, or technology—reflects the harmony of nature. In this vision:

- Renewable energy systems align with the cycles of the Earth, providing power without depletion.

- AI and robotics enhance human lives while respecting ethical boundaries.

- Circular economies flourish, transforming waste into resources and ensuring that progress does not come at the planet's expense.

By integrating \phi into the design and development of new technologies, we create a future where innovation and sustainability go hand in hand—a future where humanity's ingenuity serves as a force for harmony and renewal.

Subsection 2: Philosophies for Collective Growth and Unity

The Search for Shared Philosophies

Humanity's greatest achievements have been driven by shared philosophies—guiding principles that unite individuals, cultures, and nations in pursuit of common goals. However, modern societies face philosophical fragmentation:

1. **Conflicting Ideologies**: Divergent belief systems often lead to division and conflict, hindering collective progress.

2. **Materialism Over Meaning**: A focus on material success has overshadowed deeper questions of purpose and connection.

3. **Global Disconnection**: Despite technological advances, humanity struggles to find a unifying vision that transcends borders and cultures.

The **Golden Ratio (ϕ)** provides a universal foundation for philosophical thought, offering a framework rooted in harmony, balance, and interconnectedness. By integrating ϕ into collective philosophies, humanity can foster unity, purpose, and shared growth.

The Role of ϕ in Collective Philosophies

1. **Harmonic Values**:

   - Philosophies aligned with ϕ emphasize balance—between individual freedom and collective responsibility, progress and preservation.

   - **Example**:

     - A global charter inspired by ϕ balances environmental stewardship (62%) with economic development (38%), creating shared principles for sustainability.

2. **Universal Connection**:

   - ϕ reflects the interconnectedness of all things, inspiring philosophies that transcend cultural and national boundaries.

   - **Example**:

     - A ϕ-based education initiative integrates science, art, and spirituality to teach students about their role in the universal web of life.

3. **Proportional Growth**:

   - Philosophies guided by ϕ encourage growth that is sustainable and inclusive, fostering progress without sacrificing equity.

   - **Prediction**:

- \phi-aligned economic systems could balance technological innovation with human well-being, creating societies that thrive on shared purpose.

Applications in Collective Philosophies

1. **Education and Leadership**:

- Philosophies inspired by \phi guide curricula and leadership frameworks that prioritize holistic development and collective growth.

  - **Example**:

    - A leadership program structured with \phi-aligned modules balances technical training (62%) with emotional intelligence and ethics (38%).

2. **Cultural Integration**:

- \phi offers a unifying thread for integrating diverse cultural values into global systems, fostering mutual respect and collaboration.

  - **Example**:

    - An international summit uses \phi-inspired principles to allocate equal time to major cultural traditions (62%) and emerging perspectives (38%).

3. **Community Building**:

- Local and global communities can adopt \phi-based philosophies to create inclusive spaces that emphasize shared goals and harmonious living.

  - **Example**:

    - A community initiative uses \phi-proportioned goals to allocate resources between infrastructure improvements (62%) and cultural enrichment programs (38%).

The Benefits of \phi-Inspired Philosophies

1. **Unity Across Diversity**:

- \phi provides a common framework for reconciling diverse perspectives, fostering collaboration and mutual understanding.

2. **Sustainable Growth**:

- Philosophies aligned with \phi prioritize long-term well-being over short-term gains, ensuring equitable progress.

3. **Purpose and Connection**:

    • By reflecting the harmony of the universe, \phi-inspired philosophies offer deeper meaning and a sense of belonging.

A Vision for Unified Philosophies

Imagine a world where humanity is guided by shared philosophies that resonate with the balance and beauty of nature. In this vision:

• Education fosters a sense of interconnectedness, empowering individuals to contribute to collective growth.

• Cultures celebrate their unique identities while embracing universal principles of harmony and respect.

• Societies prioritize purpose-driven progress, balancing innovation with preservation.

By integrating \phi into the foundations of collective philosophies, humanity can transcend division and build a future of unity and shared purpose—a future where progress is measured not by individual achievement but by the harmony of the whole.

Subsection 3: A Unified Vision for Humanity's Future

The Need for a Shared Vision

As humanity faces increasingly interconnected challenges, the importance of a unified vision becomes clear. Yet, achieving this vision is often hindered by:

1. **Fragmented Goals**: Nations, organizations, and individuals frequently prioritize conflicting interests.

2. **Short-Term Thinking**: Immediate needs often overshadow long-term planning, undermining sustainability.

3. **Cultural Divides**: Differences in values and perspectives can impede global collaboration.

The **Golden Ratio (\phi)**, with its universal principles of balance, harmony, and interconnectedness, offers a foundation for crafting a shared vision that transcends these barriers. By aligning humanity's future with \phi, we can create a path forward that reflects the elegance and unity of the natural world.

Building a Unified Vision with \phi

1. **Proportional Goals**:

   - A \phi-aligned vision balances immediate needs with long-term aspirations, ensuring sustainable progress.

   - **Example**:

     - A global development plan allocates 62% of resources to addressing urgent challenges like poverty and climate change, and 38% to research and innovation for future solutions.

2. **Harmonic Systems**:

   - Systems inspired by \phi prioritize integration, creating frameworks where diverse components work together seamlessly.

   - **Example**:

     - A global transportation network uses \phi-aligned logistics to reduce environmental impact while connecting regions efficiently.

3. **Universal Collaboration**:

   - A shared vision rooted in \phi fosters unity by emphasizing common principles and values.

   - **Example**:

     - An international coalition adopts \phi-inspired policies to balance economic growth (62%) with ecological preservation (38%).

Applications of a Unified Vision

1. **Global Sustainability Initiatives**:

   - \phi-based frameworks guide efforts to balance human development with environmental stewardship.

   - **Example**:

     - A United Nations program uses \phi-aligned metrics to allocate resources proportionally across renewable energy, conservation, and education.

2. **Technology for Humanity**:

- Innovations inspired by \phi prioritize inclusivity, ensuring that technological advancements benefit all of humanity.

  - **Example**:

    - A global tech summit establishes \phi-aligned guidelines for AI development, balancing efficiency (62%) with ethical considerations (38%).

3. **Cultural Unity**:

- Cultural programs designed with \phi celebrate diversity while emphasizing shared human values.

  - **Example**:

    - An international arts festival uses \phi-spiral layouts to showcase global traditions alongside contemporary innovations.

The Benefits of a Unified Vision

1. **Shared Purpose**:

- \phi-aligned goals foster a sense of collective direction and collaboration.

2. **Sustainable Progress**:

- Balancing short- and long-term priorities ensures resilience and adaptability.

3. **Global Harmony**:

- Emphasizing interconnectedness and proportionality bridges cultural and ideological divides.

A Future Rooted in Harmony

Imagine a future where humanity moves forward as a cohesive whole—where every nation, community, and individual contributes to a shared vision of progress and harmony. In this future:

- Systems are designed to reflect the balance of nature, fostering both innovation and sustainability.

- Cultures celebrate their unique identities while collaborating on shared goals.

- Humanity's collective efforts create a world that resonates with the harmony of the cosmos.

By integrating \phi into the foundations of this vision, we ensure that humanity's future is not only innovative but also deeply connected to the timeless rhythms of the universe—a future where progress and unity go hand in hand.

Conclusion: A Future Guided by Universal Harmony

The future of humanity depends on our ability to align progress with the principles of balance, harmony, and interconnectedness. In a world facing unprecedented challenges, the **Golden Ratio (\phi)** offers a timeless and universal framework for navigating the complexities of our time. By embracing \phi as a guiding principle, we unlock a path forward that is both innovative and sustainable.

Key Insights from Section 8

1. **Sustainable Innovation**:

   - \phi-aligned advancements in technology and systems design ensure that progress supports both human needs and planetary health.

2. **Philosophies for Unity**:

   - Shared philosophies rooted in \phi foster collective growth, transcending cultural and ideological divides.

3. **A Unified Vision**:

   - By balancing immediate goals with long-term aspirations, humanity can create a future that reflects the harmony and proportionality of the cosmos.

The Benefits of a Future Aligned with \phi

1. **Equity and Inclusivity**:

   - Proportional frameworks ensure that resources and opportunities are distributed fairly, fostering global harmony.

2. **Sustainability**:

   - Systems inspired by \phi prioritize renewal and resilience, ensuring that progress does not come at the expense of future generations.

3. **Unity Across Diversity**:

   - By emphasizing interconnectedness, \phi bridges divides and fosters collaboration across cultures, nations, and disciplines.

A Call to Action

To realize this vision, humanity must:

1. **Adopt Proportional Thinking**:

   - Leaders, innovators, and communities should incorporate \phi-aligned principles into their decision-making processes.

2. **Foster Collaboration**:

   - Cross-disciplinary and cross-cultural partnerships are essential for creating systems that reflect the harmony of \phi.

3. **Inspire Generations**:

   - Education, art, and storytelling must champion the principles of \phi, instilling a sense of purpose and unity in future generations.

A Vision for the Ages

Imagine a world where humanity thrives not in opposition to nature but in harmony with it—a world where progress reflects the timeless rhythms of the universe. In this vision:

- Systems are designed with balance and renewal in mind, ensuring sustainability.

- Cultures celebrate their unique identities while contributing to a shared global purpose.

- Innovation and tradition coexist, creating a future that honors the past while embracing the possibilities of tomorrow.

By aligning humanity's trajectory with \phi, we embrace a future where progress and harmony are inseparable—a future where the beauty and balance of the cosmos are reflected in every aspect of human life.

Conclusion: The Harmony of Progress and Balance

As humanity stands at the precipice of unprecedented change, we are presented with a rare opportunity to redefine our relationship with the world, each other, and ourselves. The principles of the **Golden Ratio (\phi)**—proportionality, harmony, and interconnectedness—offer a timeless blueprint for navigating the complexities of our time.

This book has explored how ϕ can inspire and guide us across disciplines:

- In **physics and cosmology**, ϕ reveals the hidden patterns of the universe, reminding us of our place within the cosmic order.

- Through **energy systems**, it directs us toward sustainable solutions that align with nature's rhythms.

- In **societal systems**, ϕ offers pathways to equity, unity, and collective growth.

- For **personal development**, it encourages balance, resilience, and the pursuit of meaningful purpose.

- Across **disciplines**, ϕ serves as a unifying principle, fostering collaboration and innovation.

- In **ethical leadership**, it provides a foundation for transparency, inclusivity, and sustainable decision-making.

- And for the **future of humanity**, ϕ inspires a vision of progress that harmonizes technological advancements with environmental and social well-being.

A Vision of Universal Harmony

The essence of ϕ lies in its ability to unify—the mathematical constant that connects galaxies to sunflowers, architecture to music, science to art. By aligning ourselves with its principles, we create a world where harmony is not merely an aspiration but a lived reality.

This is a call to action for leaders, innovators, and communities alike: to embrace ϕ not just as a concept, but as a way of thinking, designing, and living. Together, we can transform systems, ideas, and lives into reflections of the universal balance that ϕ represents.

A Future Built on Balance

Imagine a world where energy flows sustainably, where leaders act with fairness, where innovation serves humanity, and where individuals find purpose in harmony with the cosmos. This future is not distant—it begins with a shift in mindset, a commitment to aligning our actions with the patterns of nature.

The Golden Ratio is not just a mathematical truth; it is a pathway to a future of beauty, equity, and renewal. In its rhythm, we find the potential to create a world worthy of the boundless promise of humanity.

www.ingramcontent.com/pod-product-compliance
Lightning Source LLC
Chambersburg PA
CBHW082117220526
45472CB00009B/2205